Lecture Notes in Computer Sc

Edited by G. Goos, J. Hartmanis, and J. van Leeuwen

Springer
*Berlin
Heidelberg
New York
Barcelona
Hong Kong
London
Milan
Paris
Singapore
Tokyo*

Josep Domingo-Ferrer (Ed.)

Inference Control
in Statistical Databases

From Theory to Practice

 Springer

Series Editors

Gerhard Goos, Karlsruhe University, Germany
Juris Hartmanis, Cornell University, NY, USA
Jan van Leeuwen, Utrecht University, The Netherlands

Volume Editor

Josep Domingo-Ferrer
Universitat Rovira i Virgili
Department of Computer Engineering and Mathematics
Av. Països Catalans 26, 43007 Tarragona, Spain
E-mail: jdomingo@etse.urv.es

Cataloging-in-Publication Data applied for

Die Deutsche Bibliothek - CIP-Einheitsaufnahme

Inference control in statistical databases : from theory to practice /
Josep Domingo-Ferrer (ed.). - Berlin ; Heidelberg ; New York ; Barcelona ;
Hong Kong ; London ; Milan ; Paris ; Tokyo : Springer, 2002
    (Lecture notes in computer science ; Vol. 2316)
    ISBN 3-540-43614-6

CR Subject Classification (1998): G.3, H.2.8, K.4.1, I.2.4

ISSN 0302-9743
ISBN 3-540-43614-6 Springer-Verlag Berlin Heidelberg New York

Springer-Verlag Berlin Heidelberg New York
a member of BertelsmannSpringer Science+Business Media GmbH

http://www.springer.de

© Springer-Verlag Berlin Heidelberg 2002
Printed in Germany

Typesetting: Camera-ready by author, data conversion by Christian Grosche, Hamburg
Printed on acid-free paper   SPIN 10846628   06/3142   5 4 3 2 1 0

Preface

Inference control in statistical databases (also known as statistical disclosure control, statistical disclosure limitation, or statistical confidentiality) is about finding tradeoffs to the tension between the increasing societal demand for accurate statistical data and the legal and ethical obligation to protect the privacy of individuals and enterprises which are the source of data for producing statistics. To put it bluntly, statistical agencies cannot expect to collect accurate information from individual or corporate respondents unless these feel the privacy of their responses is guaranteed.

This state-of-the-art survey covers some of the most recent work in the field of inference control in statistical databases. This topic is no longer (and probably never was) a purely statistical or operations-research issue, but is gradually entering the arena of knowledge management and artificial intelligence. To the extent that techniques used by intruders to make inferences compromising privacy increasingly draw on data mining and record linkage, inference control tends to become an integral part of computer science.

Articles in this book are revised versions of a few papers selected among those presented at the seminar "Statistical Disclosure Control: From Theory to Practice" held in Luxemburg on 13 and 14 December 2001 under the sponsorship of EUROSTAT and the European Union 5th FP project "AMRADS" (IST-2000-26125).

The book starts with an overview article which goes through the remaining 17 articles. These cover inference control for aggregate statistical data released in tabular form, inference control for microdata files, software developments, and user case studies. The article authors and myself hope that this collective work will be a reference point to both academics and official statisticians who wish to keep abreast with the latest advances in this very dynamic field.

The help of the following experts in discussing and reviewing the selected papers is gratefully acknowledged:

- Lawrence H. Cox (U. S. National Center for Health Statistics)
- Gerd Ronning (Universität Tübingen)
- Philip M. Steel (U. S. Bureau of the Census)
- William E. Winkler (U. S. Bureau of the Census)

As an organizer of the seminar from which articles in this book have evolved, I wish to emphasize that such a seminar would not have taken place without the sponsorship of EUROSTAT and the AMRADS project as well as the help and encouragement by Deo Ramprakash (AMRADS coordinator), Photis Nanopoulos, Harald Sonnberger, and John King (all from EUROSTAT). Finally, the inputs by Anco Hundepool (Statistics Netherlands and co-ordinator of the EU 5th FP project "CASC") and Francesc Sebé (Universitat Rovira i Virgili) were crucial to the success of the seminar and the book, respectively. I apologize for possible omissions.

February 2002   Josep Domingo-Ferrer

Table of Contents

Software and User Case Studies

Advances in Inference Control in Statistical Databases: An Overview

Josep Domingo-Ferrer

Universitat Rovira i Virgili
Dept. of Computer Engineering and Mathematics
Av. Països Catalans 26, E-43007 Tarragona, Catalonia, Spain
jdomingo@etse.urv.es

Abstract. Inference control in statistical databases is a discipline with several other names, such as statistical disclosure control, statistical disclosure limitation, or statistical database protection. Regardless of the name used, current work in this very active field is rooted in the work that was started on statistical database protection in the 70s and 80s. Massive production of computerized statistics by government agencies combined with an increasing social importance of individual privacy has led to a renewed interest in this topic. This is an overview of the latest research advances described in this book.

Keywords: Inference control in statistical database, Statistical disclosure control, Statistical disclosure limitation, Statistical database protection, Data security, Respondents' privacy, Official statistics.

1 Introduction

The protection of confidential data is a constant issue of concern for data collectors and especially for national statistical agencies. There are legal and ethical obligations to maintain confidentiality of respondents whose answers are used for surveys or whose administrative data are used to produce statistics. But, beyond law and ethics, there are also practical reasons for data collectors to care about confidentiality: unless respondents are convinced that their privacy is being adequately protected, they are unlikely to co-operate and supply their data for statistics to be produced on them.

The rest of this book consists of seventeen articles clustered in three parts:

1. The protection of tabular data is covered by the first five articles;
2. The protection of microdata (*i.e.* invididual respondent data) is addressed by the next seven articles;
3. Software for inference control and user case studies are reported in the last five articles.

The material in this book focuses on the latest research developments of the mathematical and computational aspects of inference control and should be regarded as an update of [6]. For a systematic approach to the topic, we

J. Domingo-Ferrer (Ed.): Inference Control in Statistical Databases, LNCS 2316, pp. 1–7, 2002.

strongly recommend [13]; for a quicker overview, [12] may also be used. All references given so far in this paragraph concentrate only on the mathematical and computational side of the topic. If a broader scope is required, [7] is a work where legal, organizational, and practical issues are covered in addition to the purely computational ones.

This overview goes through the book articles and then gives an account of related literature and other sources of information.

2   Tabular Data Protection

The first article "Cell suppression: experience and theory", by Robertson and Ethier, emphasizes that some basic points of cell suppression for table protection are not sufficiently known. While the underlying theory is well developed, sensitivity rules in use are in some cases flawed and may lead to the release of sensitive information. Another issue raised by the paper is the lack of a sound information loss measure to assess the damage inflicted to a table in terms of data utility by the use of a particular suppression pattern. The adoption of information-theoretic measures is hinted as a possible improvement.

The article "Bounds on entries in 3-dimensional contingency tables subject to given marginal totals" by Cox deals with algorithms for determining integer bounds on suppressed entries of multi-dimensional contingency tables subject to fixed marginal totals. Some heuristic algorithms are compared, and it is demonstrated that they are not exact. Consequences for statistical database query systems are discussed.

"Extending cell suppression to protect tabular data against several attackers", by Salazar, points out that attackers to confidentiality need not be just external intruders; internal attackers, *i.e.* special respondents contributing to different cell values of the table, must also be taken into account. This article describes three mathematical models for the problem of finding a cell suppression pattern minimizing information loss while ensuring protection for different sensitive cells and different intruders.

When a set of sensitive cells are suppressed from a table (primary suppressions), a set of non-sensitive cells must be suppressed as well (complementary suppressions) to prevent primary suppressions from being computable from marginal constraints. Network flows heuristics have been proposed in the past for finding the minimal complementary cell suppression pattern in tabular data protection. However, the heuristics known so far are only appropriate for two-dimensional tables. In "Network flows heuristics for complementary cell suppression: an empirical evaluation and extensions", by Castro, it is shown that network flows heuristics (namely multicommodity network flows and network flows with side constraints) can also be used to model three-dimensional, hierarchical, and linked tables.

Also related to hierarchical tables is the last article on tabular data, authored by De Wolf and entitled "HiTaS: a heuristic approach to cell suppression in hierarchical tables". A heuristic top-down approach is presented to find suppression patterns in hierarchical tables. When a table of high level is protected using cell suppression, its interior is regarded as the marginals of possibly several lower level tables, each of which is protected while keeping their marginals fixed.

3   Microdata Protection

The first three articles in this part describe methods for microdata protection:

- Article "Model based disclosure protection", by Polettini, Franconi, and Stander, argues that any microdata protection method is based on a formal reference model. Depending on the number of restrictions imposed, methods are classified as nonparametric, semiparametric or fully parametric. An imputation procedure for business microdata based on a regression model is applied to the Italian sample from the Community Innovation Survey. The utility of the released data and the protection achieved are also evaluated.
- Adding noise is a very used principle for microdata protection. In fact, results in the article by Yancey *et al.* (discussed below) show that noise addition methods can perform very well. Article "Microdata protection through noise addition", by Brand, contains an overview of noise addition algorithms. These range from simple white noise addition to complex methods which try to improve the tradeoff between data utility and data protection. Theoretical properties of the presented algorithms are discussed in Brand's article and an illustrative numerical example is given.
- Synthetic microdata generation is an attractive alternative to protection methods based on perturbing original microdata. The conceptual advantage is that, even if a record in the released data set can be linked to a record in the original data set, such a linkage is not actually a re-identification because the released record is a synthetic one and was not derived from any specific respondent. In "Sensitive microdata protection using Latin hypercube sampling technique", Dandekar, Cohen, and Kirkendall propose a method for synthetic microdata generation based on Latin hypercube sampling.

The last four articles in this part concentrate on assessing disclosure risk and information loss achieved by microdata protection methods:

- Article "Integrating file and record level disclosure risk assessment", by Elliot, deals with disclosure risk in non-perturbative microdata protection. Two methods for assessing disclosure risk at the record-level are described, one based on the special uniques method and the other on data intrusion simulation. Proposals to integrate both methods with file level risk measures are also presented.
- Article "Disclosure risk assessment in perturbative microdata protection", by Yancey, Winkler, and Creecy, presents empirical re-identification results that compare methods for microdata protection including rank swapping and additive noise. Enhanced re-identification methods based on probabilistic record linkage are used to empirically assess disclosure risk. Then the

performance of methods is measured in terms of information loss and disclosure risk. The reported results extend earlier work by Domingo-Ferrer *et al.* presented in [7].

- In "LHS-based hybrid microdata vs rank swapping and microaggregation for numeric microdata protection", Dandekar, Domingo-Ferrer, and Sebé report on another comparison of methods for microdata protection. Specifically, hybrid microdata generation as a mixture of original data and synthetic microdata is compared with rank swapping and microaggregation, which had been identified as the best performers in earlier work. Like in the previous article, the comparison considers information loss and disclosure risk, and the latter is empirically assessed using record linkage.

- Based on the metrics previously proposed to compare microdata protection methods (also called masking methods) in terms of information loss and disclosure risk, article "Post-masking optimization of the tradeoff between information loss and disclosure risk in masked microdata sets", by Sebé, Domingo-Ferrer, Mateo-Sanz, and Torra, demonstrates how to improve the performance of any microdata masking method. Post-masking optimization of the metrics can be used to have the released data set preserve as much as possible the moments of first and second order (and thus multivariate statistics) of the original data without increasing disclosure risk. The technique presented can also be used for synthetic microdata generation and can be extended to preserve all moments up to m-th order, for any m.

4   Software and User Case Studies

The first two articles in this part are related to software developments for the protection of statistical data:

- "The CASC project", by Hundepool, is an overview of the European project CASC (Computational Aspects of Statistical Confidentiality,[2]), funded by the EU 5th Framework Program. CASC can be regarded as a follow-up of the SDC project carried out under the EU 4th Framework Program. The central aim of the CASC project is to produce a new version of the Argus software for statistical disclosure control. In order to reach this practical goal, the project also includes methodological research both in tabular data and microdata protection; the research results obtained will constitute the core of the Argus improvement. Software testing by users is an important part of CASC as well.

- The first sections of the article "Tools and strategies to protect multiple tables with the GHQUAR cell suppression engine", by Gießing and Repsilber, are an introduction to the GHQUAR software for tabular data protection. The last sections of this article describe GHMITER, which is a software procedure allowing use of GHQUAR to protect sets of multiple linked tables. This software constitutes a very fast solution to protect complex sets of big tables and will be integrated in the new version of Argus developed under the CASC project.

This last part of the book concludes with three articles presenting user case studies in statistical inference control:

- "SDC in the 2000 U. S. Decennial Census", by Zayatz, describes statistical disclosure control techniques to be used for all products resulting from the 2000 U. S. Decennial Census. The discussion covers techniques for tabular data, public microdata files, and on-line query systems for tables. For tabular data, algorithms used are improvements of those used for the 1990 Decennial Census. Algorithms for public-use microdata are new in many cases and will result in less detail than was published in previous censuses. On-line table query is a new service, so the disclosure control algorithms used there are completely new ones.
- "Applications of statistical disclosure control at Statistics Netherlands", by Schulte Nordholt, reports on how Statistics Netherlands meets the requirements of statistical data protection and user service. Most users are satisfied with data protected using the Argus software: τ-Argus is used to produce safe tabular data, while μ-Argus yields publishable safe microdata. However, some researchers need more information than is released in the safe data sets output by Argus and are willing to sign the proper non-disclosure agreements. For such researchers, on-site access to unprotected data is offered by Statistics Netherlands in two secure centers.
- The last article "Empirical evidences on protecting population uniqueness at Idescat", by Urrutia and Ripoll, presents the process of disclosure control applied by Statistics Catalonia to microdata samples from census and surveys with some population uniques. Such process has been in use since 1995, and has been implemented with μ-Argus since it first became available.

5   Related Literature and Information Sources

In addition to the above referenced books [6,7,12,13], a number of other sources of information on current research in statistical inference control are available. In fact, since statistical database protection is a rapidly evolving field, the use of books should be directed to acquiring general insight on concepts and ideas, but conference proceedings, research surveys, and journal articles remain essential to gain up-to-date detailed knowledge on particular techniques and open issues.

This section contains a non-exhaustive list of research references, sorted from a historical point of view:

1970s and 1980s. The first broadly known papers and books on statistical database protection appear (*e.g.* [1,3,4,5,11]).

1990s. Eurostat produces a compendium for practitioners [10] and sponsors a number of conferences on the topic, namely the three *International Seminars on Statistical Confidentiality* (Dublin 1992 [9], Luxemburg 1994 [14], and Bled 1996 [8]) and the *Statistical Data Protection'98* conference (Lisbon 1998,[6]). While the first three events covered mathematical, legal, and organizational aspects, the Lisbon conference focused on the statistical, mathematical, and computational aspects of statistical disclosure control and data protection. The goals of those conferences were to promote research and interaction between scientists and practitioners in order to consolidate statistical disclosure control as a high-quality research discipline encompassing statistics, operations research, and computer science. In the second half of the 90s, the research project SDC was carried out under the EU 4th Framework Program; its most visible result was the first version of the Argus software. In the late 90s, other European organizations start joining the European Commission in fostering research in this field. A first example is Statistisches Bundesamt which organized in 1997 a conference for the German-speaking community. A second example is the United Nations Economic Commission for Europe, which has jointly organized with Eurostat two *Work Sessions on Statistical Data Confidentiality* (Thessaloniki 1999 [15] and Skopje 2001). Outside Europe, the U.S. Bureau of the Census and Statistics Canada have devoted considerable attention to statistical disclosure control in their conferences and symposia. In fact, well-known general conferences such as *COMPSTAT, U.S. Bureau of the Census Annual Research Conferences*, Eurostat's *ETK-NTTS* conference series, *IEEE Symposium on Security and Privacy*, etc. have hosted sessions and papers on statistical disclosure control.

2000s. In addition to the biennial *Work Sessions on Statistical Data Confidentiality* organized by UNECE and Eurostat, other research activities are being promoted by the U.S. Census Bureau, which sponsored the book [7], by the European projects CASC [2], and AMRADS (a co-sponsor of the seminar which originated this book).

As far as journals are concerned, there is not yet a monographic journal on statistical database protection. However, at least the following journals occasionally contain papers on this topic: *Research in Official Statistics, Statistica Neerlandica, Journal of Official Statistics, Journal of the American Statistical Association, ACM Transactions on Database Systems, IEEE Transactions on Software Engineering, IEEE Transactions on Knowledge and Data Engineering, Computers & Mathematics with Applications, Statistical Journal of the UNECE, Qüestiió* and *Netherlands Official Statistics.*

Acknowledgments

Special thanks go to the authors of this book and to the discussants of the seminar "Statistical Disclosure Control: From Theory to Practice" (L. Cox, G. Ronning, P. M. Steel, and W. Winkler). Their ideas were invaluable to write this overview, but I bear full responsibility for any inaccuracy, omission, or mistake that may remain.

References

1. N. R. Adam and J. C. Wortmann, "Security-control methods for statistical databases: A comparative study", *ACM Computing Surveys*, vol. 21, no. 4, pp. 515-556, 1989.
2. The CASC Project, http://neon.vb.cbs.nl/rsm/casc/menu.htm
3. T. Dalenius, "The invasion of privacy problem and statistics production. An overview", *Statistik Tidskrift*, vol. 12, pp. 213-225, 1974.
4. D. E. Denning and J. Schlörer, "A fast procedure for finding a tracker in a statistical database", *ACM Transactions on Database Systems*, vol. 5, no. 1, pp. 88-102, 1980.
5. D. E. Denning, *Cryptography and Data Security*. Reading MA: Addison-Wesley, 1982.
6. J. Domingo-Ferrer (ed.), *Statistical Data Protection*. Luxemburg: Office for Official Publications of the European Communities, 1999.
7. P. Doyle, J. Lane, J. Theeuwes, and L. Zayatz (eds.), *Confidentiality, Disclosure and Data Access*. Amsterdam: North-Holland, 2001.
8. S. Dujić and I. Tršinar (eds.), *Proceedings of the 3rd International Seminar on Statistical Confidentiality (Bled, 1996)*. Ljubljiana: Statistics Slovenia-Eurostat, 1996.
9. D. Lievesley (ed.), *Proceedings of the International Seminar on Statistical Confidentiality (Dublin, 1992)*. Luxemburg: Eurostat, 1993.
10. D. Schackis, *Manual on Disclosure Control Methods*. Luxemburg: Eurostat, 1993.
11. J. Schlörer, "Identification and retrieval of personal records from a statistical data bank", *Methods Inform. Med.*, vol. 14, no.1, pp. 7-13, 1975.
12. L. Willenborg and T. de Waal, *Statistical Disclosure Control in Practice*. New York: Springer-Verlag, 1996.
13. L. Willenborg and T. de Waal, *Elements of Statistical Disclosure Control*. New York: Springer-Verlag, 2001.
14. *Proceedings of the 2nd International Seminar on Statistical Confidentiality (Luxemburg, 1994)*. Luxemburg: Eurostat, 1995.
15. *Statistical Data Confidentiality: Proc. of the Joint Eurostat/UNECE Work Session on Statistical Data Confidentiality (Thessaloniki, 1999)*. Luxemburg: Eurostat, 1999.

Cell Suppression: Experience and Theory

Dale A. Robertson and Richard Ethier *

Statistics Canada
robedal@statcan.ca
ethiric@statcan.ca

Abstract. Cell suppression for disclosure avoidance has a well-developed theory, unfortunately not sufficiently well known. This leads to confusion and faulty practices. Poor (sometimes seriously flawed) sensitivity rules can be used while inadequate protection mechanisms may release sensitive data. The negative effects on the published information are often exaggerated. An analysis of sensitivity rules will be done and some recommendations made. Some implications of the basic protection mechanism will be explained. A discussion of the information lost from a table with suppressions will be given, with consequences for the evaluation of patterns and of suppression heuristics. For most practitioners, the application of rules to detect sensitive economic data is well understood (although the rules may not be). However, the protection of that data may be an art rather than an application of sound concepts. More misconceptions and pitfalls arise.

Keywords: Disclosure avoidance, cell sensitivity.

Cell suppression is a technique for disclosure control. It is used for additive tables, typically business data, where it is the technique of choice. There is a good theory of the technique, originally developed by Gordon Sande [1,2] with important contributions by Larry Cox [3]. In practice, the use of cell suppression is troubled by misconceptions at the most fundamental levels. The basic concept of sensitivity is confused, the mechanism of protection is often misunderstood, and an erroneous conception of information loss seems almost universal. These confusions prevent the best results from being obtained. The sheer size of the problems makes automation indispensable. Proper suppression is a subtle task and the practitioner needs a sound framework of knowledge. Problems in using the available software are often related to a lack of understanding of the foundations of the technique. Often the task is delegated to lower level staff, not properly trained, who have difficulty describing problems with the rigour needed for computer processing. This ignorance at the foundation level leads to difficulty understanding the software. As the desire for more comprehensive, detailed, and sophisticated outputs increases, the matter of table and problem specification needs further attention.

Our experience has shown that the biggest challenge has been to teach the basic ideas. The theory is not difficult to grasp, using only elementary mathematics, but clarity of thought is required. The attempt of the non-mathematical to describe things

---

* The opinions expressed in this paper are those of the authors, and not necessarily those of Statistics Canada.

J. Doningo-Ferrer (Ed.): Inference Control in Statistical Databases, LNCS 2316, pp. 8-20, 2002.
© Springer-Verlag Berlin Heidelberg 2002

in simple terms has led to confusion, and the failure to appreciate the power and value of the technique.

The idea of cell sensitivity is surprisingly poorly understood. Obsolete sensitivity rules, with consistency problems, not well adapted to automatic processing, survive. People erroneously think of sensitivity as a binary variable: a cell is publishable or confidential. The theory shows that sensitivity can be precisely measured, and the value is important in the protection process. The value helps capture some common sense notions that early practitioners had intuitively understood.

The goal of disclosure avoidance is to protect the respondents, and ensure that a response cannot be estimated accurately. Thus, a sensitive cell is one for which knowledge of the value would permit an unduly accurate estimate of the contribution of an individual respondent. Dominance is the situation where one or a small number of responses contribute nearly all the total of the cell. Identifying the two leads to the following pronouncement

"A cell should be considered sensitive if one respondent contributes 60% or more of the total value."

This is an example of an N-K rule, with N being the number of respondents to count, and K the threshold percentage. (Here N = 1 and K = 60. These values of N and K are realistic and are used in practice.)   Clearly a N-K rule measures dominance. Using it for sensitivity creates the following situation.

Consider two cells each with 3 respondents, and of total 100.
Cell 1 has a response sequence of {59,40,1} and thus may be published according to the rule. Cell 2 has a response sequence of {61,20,19}} and would be declared sensitive by the rule.

Suppose the cell value of 100 is known to the second largest respondent (X2) and he uses this information to estimate the largest (X1). He can remove his contribution to get an upper bound, obtaining (with non-negative data)

For (59,40, 1)   100-40 = 60   therefore X1 <= 60, while
For (61,20,19)   100-20 = 80   therefore X1 <= 80.

Since the actual values of X1 are 59 and 61 respectively, something has gone badly wrong. Cell 1 is much more dangerous to publish than cell 2. The "rule" gets things the wrong way around! Is this an exceptional case that took ingenuity to find? No, this problem is intrinsic to the rule, examples of misclassification are numerous, and the rule cannot be fixed. To see this we need to make a better analysis.

One can understand much about sensitivity by looking at 3 respondent cells of value 100. (Keeping the total at 100 just means that values are the same as the percentages.) Call the three response values a, b, c.

One can represent these possible cells pictorially. Recall (Fig, 1) that in an equilateral triangle, for any interior point, the sum of the perpendicular distance to the 3 sides is a constant (which is in fact h, the height of the triangle).

One can nicely incorporate the condition a + b + c = 100 by plotting the cell as a point inside a triangle, of height 100, measuring a, b, and c from a corresponding edge (Fig 2). This gives a symmetric treatment and avoids geometric confusions that occur in other plots.

In Figure 2 we have drawn the medians, intersecting at the centroid. The triangle is divided into areas which correspond to the possible size orders of a, b, and c. The upper kite shaped region is the area where a is the biggest of the three responses, with its right sub triangle the area where $a > b > c$.

In the triangle diagram, where are the sensitive cells? Cells very near an edge of the triangle should be sensitive. Near an edge one of the responses is negligible, effectively the cell has 2 respondents, hence is obviously sensitive (each respondent knows the other). As a bare minimum requirement then any rule that purports to define sensitivity must classify points near the edge as sensitive.

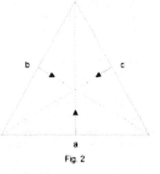

$d_1 + d_2 + d_3 = h$
Fig. 1

Fig. 2

What does the 1-60% rule do? The region where a is large is the small sub triangle at the top, away from the a edge, with its base on the line at 60% of the height of the triangle. Likewise for b and c leading to (Fig. 3)

Now one problem is obvious. Significant portions of the edges are allowed. This "rule" fails the minimum requirement. Slightly subtler is the over-suppression. That will become clearer later. Points inside the sensitive area near the interior edge and the median are unnecessarily classified as sensitive. The rule is a disaster.

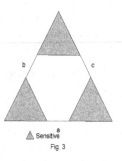

△ Sensitive
Fig. 3

Trying to strengthen the rule leads to more over-suppression, relaxing it leads to more release of sensitive data. Any organisation which defines sensitivity using only a 1-K rule (and they exist) has a serious problem. The identification of sensitivity and dominance has led to a gross error, allowing very precise knowledge of some respondents to be deduced.

Now lets look at a good rule. Here is a how a good rule divides the triangle into sensitive and non-sensitive areas. (Fig. 4)

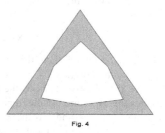

Fig. 4

The sensitivity border excludes the edges without maintaining a fixed distance from them. There is lots of symmetry and the behaviour in one of the 6 sub triangles is reproduced in all the others. This shape is not hard to understand.

This rule is expressed as a formula in terms of the responses in size order, X1, X2, and X3. As one crosses a median, the size order, and hence the expression in terms of a, b, and c changes, hence the slope discontinuities. The other thing to understand is the slope of the one non-trivial part of a sub triangle boundary, the fact that the line moves farther from the edge as one moves away from a median. The reasoning needed is a simple generalisation of the previous argument about the two cells. Start at the boundary point on the a median nearest the a axis. There a is the smallest response, while b and c are of the same size and much larger. For easy numbers suppose b=c= 45 and a=10. The protection afforded to b or c against estimation by the other comes from the presence of a. On the separation line between sensitive and non-sensitive areas, the value of the smallest response (a) is felt to be just big enough to give protection to the larger responses (b and c). The values above indicate that the large responses are allowed to be up to 4.5 times larger than the value of the response providing protection, but no larger. A higher ratio is considered sensitive, a lower one publishable. As one moves away from the median, one of b or c becomes larger than 45, the other smaller. Consequently a must become larger than 10 to maintain the critical ratio of 4.5:1. Hence a must increase (move upwards away from the bottom) i.e. the line has an upward slope.

This simple but subtle rule is one known as the C times rule. C is the ratio, an adjustable parameter (4.5 above) of the rule. (There are slight variations on this formulation, termed the p/q rule or the p% rule. They are usually specified by an inverse parameter p or $p/q = 1/C$. The differences are only in interpretation of the parameter, not in the form of the rule. We find the inverse transformation harder to grasp intuitively, and prefer this way).

The formula is easily written down then. The protection available when X2 estimates X1 is given by X3. The value of X3 has to be big enough (relative to X1) to give the needed uncertainty. One compares X1 with X3 using the scaling factor C to adjust their relative values. Explicitly one evaluates

$$S = X_1/C - X_3 \qquad (1)$$

Written in this form the rule appears not to depend on X2, but it can be trivially be written as

$$S = X_1*(C+1)/C + X_2 - T \qquad (2)$$

For a fixed total T then, the rule does depend on both X1 and X2 and they enter with different coefficients, an important point. Sensitivity depends on the cell structure (the relative sizes of the responses). This difference in coefficient values captures this structure. (Note in passing that the rule grasps the concept of sensitivity well enough that 1 and 2 respondent cells are automatically sensitive, one does not have to add those requirements as side conditions.)

The rule is trivially generalised to more than 3 respondents in the cell, X3 is simply changed to

$$X_3 + X_4 + X_5 + \ldots \qquad (3)$$

i.e. the sum of the smallest responses. One can treat all cells as effectively 3 respondent cells. (The rule can also be generalised to include coalitions where respondents in the cell agree to share their knowledge in order to estimate another respondent's contribution. The most dangerous case is X2 and X3 sharing their contributions to estimate X1. The generalisation for a coalition of 2 respondents is

$$S = X_1/C - (X_4+X_5+\ldots) \qquad (4)$$

(and similar changes for larger coalitions.) This is a deceptively simple rule of wide application, with a parameter whose meaning is clear. The C value can be taken as a precise measure of the strength of the rule.

Note that S is a function defined over the area of the triangle. Sensitivity is not just a binary classification into sensitive and non-sensitive. (S looks like an inverted pyramid with the peak at the centroid, where $a = b = c$). The line separating the two areas is the 0 valued contour of S. The value of S is meaningful. It indicates the degree of uncertainty required in estimations of the cell value, and the minimum size of an additional response big enough to render the cell non-sensitive. For example, take a 2 respondent cell of some value. Its sensitivity is $S = X1/C$. To get a non-sensitive cell S must be reduced to 0. This can be done by adding a response X3 of value $X1/C$. The cell sensitivity tells us the minimum amount that must be added to the cell to make it non-sensitive.

The only other proposed rule that occurs with any frequency is a 2-K rule; a cell is sensitive if 2 respondents contribute more than K % of the cell value. (Here K in the range 80 - 95 is typical.) Only the sum of the top two responses (X1+X2) enters. Their relative size (for a fixed total) is not used in any way. However, as just seen, the relative sizes are important. We can conclude that this cannot be a very good rule without further analysis. A 2-K rule cannot distinguish between a cell with responses (50,45,5) and one with responses (90,5,5). Most would agree that the second one is more sensitive than the first. The picture of the rule is easy to draw. If the sum of two responses is to be less than a certain value, then the third must be larger than a corresponding value. That gives a line parallel to an edge i.e. Figure 5.

As we have just observed, in a good rule the sensitivity line should not be parallel to the edge. This means that the 2-K rule is not consistent in its level of protection. In certain areas less protection is given than in others. This would seem undesirable. The rule has no clear objective. It might be hoped that the non-uniformity is negligible, and that it doesn't matter all that much. Alas no. We can quantify the non-uniformity. It turns out to be unpleasantly large. The effect of the rule is variable, and it is difficult to choose a value for K and to explain its meaning.

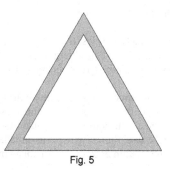

Fig. 5

To measure the non uniformity one can find the two C times rules that "surround" the 2-K rule, i.e. the strongest C times rule that is everywhere weaker than the 2-K rule, and the weakest C times rule that is everywhere stronger than the 2-K rule. These C values are easily found to be

$$C \text{ (inner)} = k/(200-k) \qquad \text{(stronger)} \qquad \qquad (5)$$

$$C \text{ (outer)} = (100-k)/k \quad \text{(weaker)} \qquad \qquad (6)$$

For the typical range of K values, this gives the following graph (Figure 6).

One can see that the difference between the C values is rather large. One can get a better picture by using a logarithmic scale (Figure 7).

On this graph, the difference between the lines is approximately constant, which means the two values differ by a constant factor. The scale shows this constant to be near 2 (more precisely it is in the range 1.7 to 1.9) i.e. the non-uniformity is serious. The problem of deciding on an appropriate value of K is ill defined.

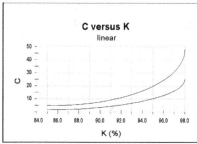

Fig. 6

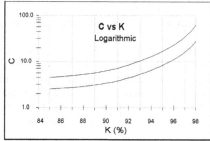

Fig. 7

The summary is then:

Don't use N-K rules, they are at best inferior with consistency problems, at worst lead to disasters. Their strength is unclear, and the K value hard to interpret. They arise from a fundamental confusion between sensitivity and dominance.

Now its time to talk a little about the protection mechanism. First, observe that in practice, the respondents are businesses. Rough estimates of a business size are easy to make. In addition, the quantities tabulated are often intrinsically non-negative. These two things will be assumed; suppressed cells will only be allowed to take values within 50% of the true ones from now on

Here is a trivial table with 4 suppressions X1, X2, X3, X4 (Figure 8).

One or more of these cells may be presumed sensitive, and the suppressions protect the sensitive values from trivial disclosure. Suppose X1 is sensitive. Note that X1+X2 and X1+X3 are effectively published. The suppressed data are not lost, they are simply aggregated. Aggregation is the protection mechanism at work here, just as in statistics in general. Data is neither lost nor distorted.

103	101	2
101	**X1**	**X2**
2	**X3**	**X4**

Fig. 8

Suppression involves the creation of miscellaneous aggregations of the sensitive cells with other cells. Obviously then if the pattern is to be satisfactory, then these aggregations must be non-sensitive. Here the unions X1+X2 and X1+X3 must be non-sensitive if this pattern is to be satisfactory. From our previous discussion if follows that both X2 and X3 must be at least as large as the sensitivity of X1. As complements, the best they can do is to add value to the protective responses. There should be no responses from the large contributors in the complementary cells. Proper behaviour of S when cells are combined will be of great importance. Certainly one would like to ensure as a bare minimum

Two non-sensitive cells should never combine to form a sensitive cell.

The notion that aggregation provides protection is captured by the property of sub-additivity (1), (2). This is an inequality relating the sensitivity of a combination of two cells to the sensitivity of the components. The direction of the inequality is such that aggregation is a good thing. One should also have smoothness of behaviour, the effect on sensitivity should be limited by the size of the cell being added in. For a good rule then one has the inequalities

$$S(x) - T(Y) <= S(X+Y) <= S(X) + S(Y) \tag{7}$$

(Using a convenient normalisation for S)

Given that S >= 0 indicates sensitivity, and that the aim is to ensure that S(X+Y) is not sensitive given that S(X) is, the right most inequality indicates that aggregation tends to be helpful (and is definitely helpful if the complement Y is not sensitive, S(Y) < 0).   The left inequality limits the decrease in the S value. A successful complement must be big enough, T (Y) >= S(X) to allow S(X+Y) to be negative. These inequalities are natural conditions that most simple rules obey, but people have, at some effort, found rules that violate them.

These inequalities are only that. One cannot exactly know the sensitivity of a union by looking at the summary properties of the cells. One needs the actual responses. The typical complicating factor here is a respondent who has data in more than one cell of a union. (This may be poor understanding of the corporate structure or data classified at an unsuitable level.) Generation of a good pattern is thus a subtle process, involving the magnitudes and sensitivities of the cells, and the pattern of individual responses in the cells.   One needs information about the individual responses when creating a pattern. The sensitive union problem certainly is one that is impractical to perform without automation.   Sophisticated software is indispensable. Tables that appear identical may need different suppression patterns, because of the different underlying responses.   We have found, using a small pedagogical puzzle, that few people understand this point

In the software in use at Statistics Canada (CONFID) the sensitive union problem is dealt with by enumerating all sensitive combinations which are contained in a non-sensitive totals [4].   (They are in danger of being trivially disclosed.) All such combinations are protected by the system.   This may sound like combinatorial explosion and a lot of work. It turns out that there are not that many to evaluate, the extra work is negligible, and the effect on the pattern small. The algorithm makes heavy use of the fact that non-sensitive cells cannot combine to form a sensitive union.

Now let us talk a bit about the suppression process, and information.   It is generally said that one should minimise the information lost, but without much discussion of what that means.   It is often suggested that the information lost in a table with suppressions can be measured by

> i) the number of suppressed cells,
> ii) the total suppressed value.

The fact that one proposes to measure something in two different and incompatible ways surely shows that we are in trouble.   Here are two tables with suppressions (Figure 9, Figure 10).

103	101	2
101	X	X
2	X	X

Fig. 9

103	77	26
77	X	X
26	X	X

Fig. 10

These two tables have the same total number of suppressed cells, 4, and the same total suppressed value, 103. By either of the above incompatible definitions then, they should have the same amount of missing information. However, they do not. Substantially more information has been removed from one of these two tables than from the other (three times as much in fact). This misunderstanding about information is related to the misunderstanding that the value of a suppressed cell remains a total mystery, with no guesses about the value possible, and that somehow the data is lost. In fact, it is just aggregated. Furthermore, any table with suppressions is equivalent to a table of ranges for the hidden cells. This latter fact is not well known, or if it is, the implications are not understood. These ranges can be found by a straightforward and inexpensive LP calculation, maximising and minimising the value of each hidden cell subject to the constraint equations implied by the additivity of the table. The above tables are equivalent to Figure.11, Figur 12, (provided the data are known to be non-negative). The different ranges provide a clue about the information loss. The second table has wider ranges. Large complementary cells used to protect cells of small sensitivity often turn out to have very narrow ranges and contain valuable information. They should not be thought of as lost.

103	101	2
101	99-101	0-2
2	0-2	0-2

Fig. 11

103	77	26
77	51-77	0-26
26	0-26	0-26

Fig. 12

Clearly one needs a better concept of information. Looking around, in the introductory chapters of an introductory book we found (in the context of signal transmission in the presence of errors) a thought that we paraphrase as

The additional information required to correct or repair a noisy signal (thus recovering the original one) is equal to the amount of information which has been lost due to the noise.

Thinking of a table with missing entries as a garbled message, the information missing from a table is the minimum amount of information needed to regain the full table. This may seem simple, but it is a subtle concept. One can use all properties that are implied by the table structure. The previous measures do not use the structural properties. The width of the ranges will have some bearing. One also can use the fact that the hidden values are not independent, but are linked by simple equations. Only a subset of them need be recovered by using more information, the rest can be trivially calculated. Calculating the cost on a cell by cell basis implicitly suggests that all hidden cells are independent, and that the information loss can simply be added up.

For the first table (assuming all the quantities are integral for simplicity) there are only 3 possible tables compatible with the suppression pattern. Consequently, (having agreed upon a standard order for enumerating the possible tables) one only needs to be told which table in the sequence is the true one, table 1, table 2 or table 3. In this case the amount of information needed is that needed to select one out of three (equally probable) possibilities. This is a precise amount of information, which even has a name, one trit. In more common units 1 trit = 1.58 bits (since the binary log of 3 is 1.58...). For the second table one has 27 possible solutions. Selecting one from 27 could be done by 3 divisions into 3 equal parts selecting one part after each division. So one requires 3 trits of information, hence the statement that 3 times as much information has been lost in the second table. Counting the number of tables correctly uses the cell ranges, and the relationships between the hidden values, both of which are ignored by the value or number criteria.

This viewpoint can resolve some other puzzles or paradoxes. Here is a hypothetical table in which the two cells of value 120 are sensitive (Figure 13).

(120)	10	20	0	100
10	10	0	0	0
20	0	0	0	20
0	0	0	10	10
100	0	20	10	(120)

○ Sensitive
● Total

Fig. 13

Here is the minimum cell count suppression pattern, with two complements totalling 200, and the minimum suppressed value pattern, (Figure 14, Figure 15), with 6 complements totalling 60.

X	10	20	0	X
10	10	0	0	0
20	0	0	0	20
0	0	0	10	10
X	0	20	10	X

● Total

Minimum Count Pattern

Fig. 14

X	X	20	0	100
X	X	0	0	0
20	0	0	0	20
0	0	0	X	X
100	0	20	X	X

● Total

Minimum Value Pattern

Fig. 15

Here (Figure 16) is the pattern we prefer, which is intermediate between the two others, having 4 complementary suppressions of total value 80.

X	10	X		0	100
10	10	0		0	0
X	0	0		0	X
0	0	0		10	10
100	0	X		10	X

Total

Recommended Pattern

Fig. 16

Using the notion of the amount of information needed to recover the table, this is in fact the best of the 3.   With the minimum count pattern, the size of the complements makes the ranges large, and there are many possible tables (101) consistent with the pattern.   With the minimum value pattern, although the ranges are small, there are two independent sets of hidden cells, and hence the number of tables is the product of the numbers of the two sub-tables with suppression (11*11). In the preferred pattern one has 21 possible tables.

(Note for the minimum value pattern, one has two independent sub-patterns. Here one would like to be able to say that the information lost is the sum of two terms, one for each independent unit. Since the number of tables is the product of the numbers of possible solutions to these two sub-problems, it is clear that it is appropriate to take logarithms of the number of possible tables. Some of you may of course realise that this is a vast oversimplification of information theory.   There is a simple generalisation of the measure if the tables are not equi-probable.) The form of objective function that is in practice the most successful in CONFID may be looked on as an attempt to approximate this sort of measure. Given all these subtleties, it follows that the effects of interventions, forcing the publication or suppression of certain cells should not be done without serious thought. One should always measure the effect of this type of intervention.

Given that tables with suppressions are equivalent to range tables, and that sophisticated users are probably calculating these ranges themselves, either exactly or approximately, it has often been suggested that the statistical agencies improve their service, especially to the less sophisticated users by publishing the ranges themselves.   One suppression package ACS [5] takes this farther by providing in addition to the ranges, a hypothetical solution consistent with them, i.e. a conceivable set of values for the hidden cells which make the tables add up. These values are not an estimate in any statistical sense, but provide a full set of reasonable values that may be of use in certain types of models for example, which do not like missing data points.   In our view, range publication would be highly desirable.   It has the following advantages for the statistical agency.

The personnel doing the suppression need to think more and to have a better understanding of the process. Seeing the ranges will add to the quality of the patterns, especially if any hand tinkering has happened. The effect of tinkering can be better evaluated.

Certain „problems" attributed to cell suppression are pseudo-problems, caused only by poor presentation. One such is the feared necessity of having to use a large complement to protect a not very sensitive cell. Well if the sensitivity is small, the cell doesn't require much ambiguity or protection. Most of the statistical value of the complementary cell can be published.

1995	1996	1997	1998	1999	2000
72	74	X	69	X	70

Fig. 17

1995	1996	1997	1998	1999	2000
72	74	69-74	69	67-73	70

Fig. 18

Another problem, the issue of continuity in time series, becomes smaller in importance. It is generally felt that a series such as Figure 17 is disconcerting. If ranges were used one could publish something like Figure 18 which is less disturbing.

Obviously there are advantages for the customer too. They are getting more data. (Or the data more conveniently. More sophisticated users can perform the analysis for themselves.) Giving the ranges explicitly helps the less sophisticated user.

In our opinion, the arguments for range publication are convincing, and any objections are not very sensible. If one had competing statistical agencies, they would be rushing to get this new and improved service out to the salesman.

A few conclusions

If we don't yet have a good measure of information lost in a table, it follows that all methods in use today are heuristic. They solve various problems that approximate the real problem. It is difficult to quantify how well these approximate problems resemble the real problem. Our feeling is that better progress would be attained if one had agreed upon the proper problem, and discussed various methods of solution, exact or approximate. Properties of the problem and its solution could be studied, and a more objective way to evaluate the heuristic methods would be available. As well, standard test data sets could be prepared to facilitate discussion and comparison.

It is only a modest overstatement to suggest that some people don't know what they are doing, in spite of the fact that a proper understanding is not difficult. Therefore training and attitude are big problems. The common errors that occur are using bad, deeply flawed sensitivity rules, and using inadequate methods to generate the patterns that do not ensure that the sensitive cells have sufficient ambiguity or protect combinations.

References

[1] Towards Automated Disclosure Analysis for Statistical Agencies.   Gordon Sande; InternalDocument Statistics Canada (1977)
[2] Automated Cell Suppression to Preserve Confidentiality of Business Statistics. Gordon Sande; Stat. Jour U.N. ECE2 pp33-41 (1984)
[3] Linear Sensitivity Measures in Statistical disclosure Control. L.H. Cox.; Jour. Stat. Plan. & Infer. V5, pp153-164 (1981)
[4] Improving Statistics Canada's Cell Suppression Software (CONFID). D. A. Robertson, COMPSTAT 2000 Proceedings in Computational Statistics. Ed. J.K Bethlehem, P.G.M van der Heiden, Physica Verlag (Heidelberg New York) (2000)
[5] ACS available from Sande and Associates, 600 Sanderling Ct., Secaucus N.J. 07094 U.S.A.   g.sande@worldnet.att.net

Bounds on Entries in 3-Dimensional Contingency Tables Subject to Given Marginal Totals

Lawrence H. Cox

U.S. National Center for Health Statistics, 6525 Belcrest Road
Hyattsville, MD 20782 USA
lcox@cdc.gov

Abstract: Problems in statistical data security have led to interest in determining exact integer bounds on entries in multi-dimensional contingency tables subject to fixed marginal totals. We investigate the 3-dimensional integer planar transportation problem (3-DIPTP). Heuristic algorithms for bounding entries in 3-DIPTPs have recently appeared. We demonstrate these algorithms are not exact, are based on necessary but not sufficient conditions to solve 3-DIPTP, and that all are insensitive to whether a feasible table exists. We compare the algorithms and demonstrate that one is superior, but not original. We exhibit fractional extremal points and discuss implications for statistical data base query systems.

1   Introduction

A problem of interest in operations research since the 1950s [1] and during the 1960s and 1970s [2] is to establish sufficient conditions for existence of a feasible solution to the *3-dimensional planar transportation problem (3-DPTP)*, viz., to the linear program:

$$\sum_{i=1}^{d_1} n_{ijk} = n_{+jk}, \quad \sum_{j=1}^{d_2} n_{ijk} = n_{i+k}, \quad \sum_{k=1}^{d_3} n_{ijk} = n_{ij+}, \quad n_{ijk} \geq 0, \tag{1}$$

where n_{+jk}, n_{i+k}, $n_{ij+} \geq 0$ are constants, referred to as the *2-dimensional marginal totals*. Attempts on the problem are summarized in [3]. Unfortunately, each has been shown [2, 3] to yield necessary but not sufficient conditions for feasibility. In [4] is given a sufficient condition for multi-dimensional transportation problems based on an iterative nonlinear statistical procedure known as *iterative proportional fitting*. The purpose here is to examine the role of feasibility in the pursuit of exact integral lower and upper bounds on internal entries in a 3-DPTP subject to integer constraints (viz., a *3-DIPTP*) and to describe further research directions. Our notation suggests that internal and marginal entries are integer. Integrality is not required for 3-DPTP (1), nor by the feasibility procedure of [4]. However, henceforth integrality of all entries is assumed as we focus on *contingency tables* - tables of nonnegative integer frequency counts and totals - and on the 3-DIPTP.

J. Domingo-Ferrer (Ed.): Inference Control in Statistical Databases, LNCS 2316, pp. 21 - 33, 2002.

The *consistency conditions* are necessary for feasibility of 3-DIPTP:

$$\sum_{k=1}^{d_3} n_{i+k} = \sum_{j=1}^{d_2} n_{ij+}, \ \sum_{i=1}^{d_1} n_{ij+} = \sum_{k=1}^{d_3} n_{+jk}, \ \sum_{i=1}^{d_1} n_{i+k} = \sum_{j=1}^{d_2} n_{+jk}, \ n_{ijk} \geq 0. \quad (2)$$

The respective values, n_{i++} , n_{+j+} , n_{++k}, are the *1-dimensional marginal totals*.

$$n_{+++} = \sum_{i=1}^{d_1} n_{i++} = \sum_{j=1}^{d_2} n_{+j+} = \sum_{k=1}^{d_3} n_{++k} \text{ is the } \textit{grand total}. \text{ It is customary to}$$

represent the *2-dimensional marginal totals* in matrices defined by elements:

$$A_{jk} = \sum_{i=1}^{d_1} n_{ijk} , \ B_{ik} = \sum_{j=1}^{d_2} n_{ijk} , \ C_{ij} = \sum_{k=1}^{d_3} n_{ijk}, \quad n_{ijk} \geq 0. \quad (3)$$

The *feasibility problem* is the existence of integer solutions to (1) subject to consistency (2) and integrality conditions on the 2-dimensional marginal totals. The *bounding problem* is to determine integer lower and upper bounds on each entry n_{ijk} over contingency tables satisfying (1)-(2). *Exact bounding* determines the interval $[\min\{n_{ijk}\}, \max\{n_{ijk}\}]$, over all integer feasible solutions $n^* = \{n^*_{ijk}\}$ of (1)-(2).

The bounding problem is important in statistical data protection. To prevent unauthorized disclosure of confidential subject-level data, it is necessary to thwart narrow estimation of small counts. In lieu of releasing the internal entries of a 3-dimensional contingency table, a national statistical office (NSO) may release only the 2-dimensional marginals. An important question for the NSO is then: how closely can a third party estimate the suppressed internal entries using the published marginal totals? During large-scale data production such as for a national census or survey, the NSO needs to answer this question thousands of times.

Several factors can produce an *infeasible table*, viz., marginal totals satisfying (1)-(2) for which no feasible solution exists, and that infeasible tables are *ubiquitous* and *abundant*, viz., dense in the set of all potential tables [4]. To be useful, bounding methods must be sensitive to infeasibility, otherwise meaningless data and erroneous inferences can result [5].

The advent of public access statistical data base query systems has stimulated recent research by statisticians on the bounding problem. Unfortunately, feasibility and its consequences have been ignored. We highlight and explore this issue, through examination of four papers representing separate approaches to bounding problems. Three of the papers [6-8] were presented at the International Conference on Statistical Data Protection (*SDP'98*), March 25-27, 1998, Lisbon, Portugal, sponsored by the Statistical Office of the European Communities (EUROSTAT). The fourth [9] appeared in *Management Science* and reports current research findings. A fifth, more recent, paper [10] is also discussed. We examine issues raised and offer observations and generalizations.

2   The F-Bounds

Given a 2-dimensional table with consistent sets of column ($\{n_{+j}\}$) and row ($\{n_{i+}\}$) marginal totals, the *nominal upper bound* for n_{ij} equals $\min\ \{n_{+j},\ n_{i+}\}$. The *nominal lower bound* is zero.

It is easy to obtain exact bounds in 2-dimensions. The nominal upper bound is exact, by the *stepping stones algorithm*: set n_{ij} to its nominal upper bound, and subtract this value from the column, row and grand totals. Either the column total or the row total (or both) must become zero: set all entries in the corresponding column (or row, or both) equal to zero and drop this column (or row, or both) from the table. Arbitrarily pick an entry from the remaining table, set it equal to its nominal upper bound, and continue. In a finite number of iterations, a completely specified, consistent 2-dimensional table exhibiting the nominal upper bound for n_{ij} will be reached. Exact lower bounds can be obtained as follows.

$$\text{As } n_{+j} + n_{i+} - n_{++} = n_{ij} - \sum_{I \neq i,\ J \neq j} n_{IJ}\ , \text{then}$$

$n_{ij} \geq \max\ \{0,\ n_{+j} + n_{i+} - n_{++}\}$. That this bound is exact follows from observing that $\sum_{I \neq i,\ J \neq j} n_{IJ} = 0$ is feasible if

$n_{+j} + n_{i+} - n_{++} \geq 0$. Therefore, in 2-dimensions, exact bounds are given by:

$$\min\ \{n_{+j},\ n_{i+}\} \geq n_{ij} \geq \max\ \{0,\ n_{+j} + n_{i+} - n_{++}\}. \tag{4}$$

These bounds generalize to m-dimensions, viz., each internal entry is contained in precisely m(m-1)/2 2-dimensional tables, each of which yields a candidate lower bound. The maximum of these lower bounds and zero provides a lower bound on the entry. Unlike the 2-dimensional case, in $m \geq 3$ dimensions these bounds are not necessarily exact [5]. We refer to these bounds as the *F-bounds*. In 3-dimensions, the F-bounds are:

$$\min\ \{n_{+jk},\ n_{i+k},\ n_{ij+}\} \geq n_{ijk}$$
$$\geq \max\ \{0,\ n_{ij+} + n_{i+k} - n_{i++},\ n_{ij+} + n_{+jk} - n_{+j+},\ n_{i+k} + n_{+jk} - n_{++k}\}. \tag{5}$$

3   The Bounding Methods of Fienberg and Buzzigoli-Giusti

3.1   The Procedure of Fienberg

In [6], Fienberg does not specify a bounding algorithm precisely, but illustrates an approach via example. The example is a 3x3x2 table of sample counts from the 1990 Decennial Census Public Use Sample [6, Table 1]. For convenience, we present it in the following form. The internal entries are:

Table 1. Fienberg [6, Table 1]

INCOME

	High Med Low	High Med Low
White	96 72 161	186 127 51
Black	10 7 6	11 7 3
Chinese	1 1 2	0 1 0
	MALE	FEMALE

and the 2-dimensional marginal totals are:

$$
A = \begin{pmatrix} 107 & 197 \\ 80 & 135 \\ 169 & 54 \end{pmatrix}, \; B = \begin{pmatrix} 329 & 364 \\ 23 & 21 \\ 4 & 1 \end{pmatrix}, \; C = \begin{pmatrix} 282 & 199 & 212 \\ 21 & 14 & 9 \\ 1 & 2 & 2 \end{pmatrix}.
$$

In the remainder of this sub-section, we examine the properties of the bound procedure of [6].

Corresponding to internal entry n_{ijk} , there exists a *collapsed* 2x2x2 table:

Table 2. Collapsing a 3-dimensional table around entry n_{ijk}

n_{ijk}	$\sum_{J \neq j} n_{iJk}$	$\sum_{K \neq k} n_{ijK}$	$\sum_{J \neq j, K \neq k} n_{iJK}$
$\sum_{I \neq i} n_{Ijk}$	$\sum_{I \neq i, J \neq j} n_{IJk}$	$\sum_{I \neq i, K \neq k} n_{IjK}$	$\sum_{I \neq i, J \neq j, K \neq k} n_{IJK}$

Entry (2, 2, 2) in the lower right-hand corner is the *complement* of n_{ijk}, denoted $\bar{n}_{ijk}$. Fix (i, j, k). Denote the 2-dimensional marginals of Table 2 to which n_{ijk} contributes by $\bar{n}_{+22}, \bar{n}_{2+2}, \bar{n}_{22+}$. Observe that:

$$
\begin{aligned}
& n_{+++} - n_{i++} - n_{+j+} - n_{++k} + n_{ij+} + n_{i+k} + n_{+jk} - \bar{n}_{ijk} \\
&= (n_{+++} - n_{i++} - n_{+j+} + n_{ij+}) + (n_{i+k} + n_{+jk} - n_{++k}) - (\bar{n}_{ijk}) \\
&= (\bar{n}_{22+}) + (n_{ijk} - (\bar{n}_{22+} - \bar{n}_{ijk})) - (\bar{n}_{ijk}) = n_{ijk}.
\end{aligned}
\tag{6}
$$

From (6) results the *2-dimensional Fréchet lower bound* of Fienberg (1999):

$$
n_{ijk} \geq \max \{0, n_{+++} - n_{i++} - n_{+j+} - n_{++k} + n_{ij+} + n_{i+k} + n_{+jk} - \min\{\bar{n}_{+22}, \bar{n}_{2+2}, \bar{n}_{22+}\}\}.
\tag{7}
$$

The nominal (also called *Fréchet*) upper bound on n_{ijk} equals $\min \{n_{+jk}, n_{i+k}, n_{ij+}\}$. The *2-dimensional Fréchet bounds* of [6] are thus:

$$\min \{n_{+jk}, n_{i+k}, n_{ij+}\} \geq n_{ijk} \geq$$
$$\max\{0, n_{+++} - n_{i++} - n_{+j+} - n_{++k} + n_{ij+} + n_{i+k} + n_{+jk} - \min \{\bar{n}_{+22}, \bar{n}_{2+2}, \bar{n}_{22+}\}\} \ . \tag{8}$$

Simple algebra reveals that the lower F-bounds of Section 2 and the 2-dimensional lower Fréchet bounds of [6] are identical. The F-bounds are easier to implement. Consequently, we replace (8) by (5).

From (6) also follows the *2-dimensional Bonferroni upper bound* of [6]:

$$n_{+++} - n_{i++} - n_{+j+} - n_{++k} + n_{ij+} + n_{i+k} + n_{+jk} \geq n_{ijk} \ . \tag{9}$$

This Bonferroni upper bound is not redundant: if $\max \{\bar{n}_{ijk}\}$ is sufficiently small, it can be sharper than the nominal upper bound. This is illustrated by the n_{111} entry of the 2x2x2 table with marginals:

$$A = \begin{pmatrix} 2 & 8 \\ 9 & 8 \end{pmatrix}, \quad B = \begin{pmatrix} 2 & 9 \\ 9 & 7 \end{pmatrix}, \quad C = \begin{pmatrix} 2 & 9 \\ 8 & 8 \end{pmatrix} \tag{10}$$

$n_{i++} + n_{+j+} + n_{++k} - 2n_{+++} \leq n_{ijk} - \bar{n}_{ijk}$ yields the *1-dimensional Fréchet lower bound* [6]:

$$n_{ijk} \geq n_{i++} + n_{+j+} + n_{++k} - 2n_{+++} \tag{11}$$

This bound is redundant with respect to the lower F-bounds. This is demonstrated as follows:

$$1/3[(n_{ij+} + n_{i+k} - n_{i++}) + (n_{ij+} + n_{+jk} - n_{+j+}) + (n_{i+k} + n_{+jk} - n_{++k})]$$
$$- (n_{i++} + n_{+j+} + n_{++k} - 2n_{+++}) \tag{12}$$
$$= 2/3(\bar{n}_{+22} + \bar{n}_{2+2} + \bar{n}_{22+}) \geq 0.$$

The *Fréchet-Bonferroni bounds* of [6] can be replaced by:

$$\min \{n_{+++} - n_{i++} - n_{+j+} - n_{++k} + n_{ij+} + n_{i+k} + n_{+jk}, n_{+jk}, n_{i+k}, n_{ij+}\} \geq n_{ijk} \geq$$
$$\max \{0, n_{ij+} + n_{i+k} - n_{i++}, n_{ij+} + n_{+jk} - n_{+j+}, n_{i+k} + n_{+jk} - n_{++k}\} \ . \tag{13}$$

We return to the example [6, Table 1]. Here, the 2-dimensional Bonferroni upper bound (9) is not effective for any entry, and can be ignored. Thus, in this example, the bounds (13) are identical to the F-bounds (5), and should yield identical results. They in fact do not, most likely due to numeric error somewhere in [6]. This discrepancy is need to be kept in mind as we compare computational approaches below, that of Fienberg, using (13), and an alternative using (5).

Fienberg [6] computes the Fréchet bounds, without using the Bonferroni bound (9), resulting in Table 7 of [6]. We applied the F-bounds (5), but in place of his Table 7, obtained sharper bounds. Both sets of bounds are presented in Table 3, as follows. If our bound agrees with [6 , Table 7] we present the bound. If there is disagreement, we include the [6], Table 7 bound in parentheses alongside ours.

Table 3. F-Bounds and Fienberg (Table 7) Fréchet Bounds for Table 1

INCOME

High	Med	Low		High	Med	Low
(80)85	(53)64	(142)158		175	(113)119	(32)43
−107	−80	−169		−197	−135	−54
0−21	0−14	0−9		0−21	0−14	0−9
0−1	0−2	0−2		0−1	0−1	0−1
	MALE				FEMALE	

Fienberg [6] next applies the Bonferroni upper bound (9) to Table 7, and reports improvement in five cells and a Table 8, the table of exact bounds for the table. We were unable to reproduce these results: the Bonferroni bound provides improvement for none of the entries.

3.2   The Shuttle Algorithm of Buzzigoli-Giusti

In [7], Buzzigoli-Giusti present the iterative *shuttle algorithm*, based on *principles of subadditivity*:
- a sum of lower bounds on entries is a lower bound for the sum of the
        entries, and
- a sum of upper bounds on entries is an upper bound for the sum of the
        entries;
- the difference between the value (or an upper bound on the value) of an
        aggregate and a lower bound on the sum of all but one entry in the
        aggregate is an upper bound for that entry, and
- the difference between the value (or a lower bound on the value) of an
        aggregate and an upper bound on the sum of all but one entry in
        the aggregate is a lower bound for that entry.

The shuttle algorithm begins with nominal lower and upper bounds. For each entry and its 2-dimensional marginal totals, the sum of current upper bounds of all other entries contained in the 2-dimensional marginal is subtracted from the marginal. This is a candidate lower bound for the entry. If the candidate improves the current lower bound, it replaces it. This is followed by an analogous procedure using sums of lower bounds and potentially improved upper bounds. This two-step procedure is repeated until all bounds are stationary. The authors fail to note but it is evident that *stationarity* is reached in a finite number of iterations because the marginals are integer.

3.3   Comparative Analysis of Fienberg , Shuttle, and F-Bounding Methods

We compare the procedure of Fienberg ([6]), the shuttle algorithm and the F-bounds. As previously observed, the bounds of [6] can be reduced to the F-bounds plus the 2-dimensional Bonferroni upper bound, viz., (13).   The shuttle algorithm produces bounds at least as sharp as the F-bounds, for two reasons.   First, the iterative shuttle procedure enables improved lower bounds to improve the nominal and subsequent upper bounds.   Second, lower F-bounds can be no sharper than those produced during the first set of steps of the shuttle algorithm.   To see this, for concreteness consider the candidate lower F-bound $n_{i+k} + n_{+jk} - n_{++k}$ for n_{ijk}. One of the three candidate shuttle lower bounds for n_{ijk} equals $n_{i+k} - \sum_{J \neq j} n^{U}_{iJk}$, where n^{U}_{iJk} denotes the current upper bound for n_{iJk}.

Thus, $n^{u}_{iJk} \leq n_{+Jk}$ and therefore $\sum_{J \neq j} n^{U}_{+Jk} \leq \sum_{J \neq j} n_{+Jk} = n_{++k} - n_{+jk}$.

Consequently, the shuttle candidate lower bound is greater than or equal to

$n_{i+k} - (n_{++k} - n_{+jk}) = n_{i+k} + n_{+jk} - n_{++k}$, so the shuttle candidate is at least as sharp as the F-candidate.   If the shuttle algorithm is employed, all but the nominal lower Fienberg (1999) and F-bounds (namely, 0) are redundant.

Buzzigoli-Giusti [7] illustrate the 3-dimensional shuttle algorithm for the case of a 2x2x2 table.   It is not clear, for the general case of a $d_1 x d_2 x d_3$ table, if they intend to utilize the collapsing procedure of Table 2, but in what follows we assume that they do. Consider the 2-dimensional Bonferroni upper bounds (9).   From (6), the Bonferroni upper bound for n_{ijk} equals $n_{ijk} + \bar{n}_{ijk}$.   Consider the right-hand 2-dimensional table in Table 2.   Apply the standard 2-dimensional lower F-bound to the entry in the upper left-hand corner:

$$\sum_{K \neq k} n_{ijK} \geq \sum_{K \neq k} n_{ijK} - \bar{n}_{ijk} = ((n_{i++} - n_{i+k}) + (n_{+j+} - n_{+jk}) - (n_{+++} - n_{++k})) .   \quad (14)$$

As previously observed, the shuttle algorithm will compute this bound during step 1 and, if it is positive, replace the nominal lower bound with it, or with something sharper. During step 2, the shuttle algorithm will use this lower bound (or something sharper) to improve the upper bound on n_{ijk}, as follows:

$$n_{ijk} = n_{ij+} - \sum_{k \neq K} n_{ijK} \leq n_{ij+} - ((n_{i++} - n_{i+k}) + (n_{+j+} - n_{+jk}) - (n_{+++} - n_{++k}))$$
$$= n_{+++} - n_{i++} - n_{+j+} - n_{++k} + n_{ij+} + n_{i+k} + n_{+jk} . \quad (15)$$

Thus, the Fienberg [6] 2-dimensional Bonferroni upper bound is redundant relative to the shuttle algorithm.   Consequently, if the shuttle and collapsing methodologies are applied in combination, it suffices to begin with the nominal bounds and run the shuttle algorithm to convergence.   Application of this approach (steps 1-2-1) to Table 1 yields Table 4 of exact bounds:

Table 4. Exact bounds for Table 1 from the Shuttle Algorithm

INCOME

High	Med	Low		High	Med	Low
85-107	64-79	158-168		175-197	120-135	44-54
0-21	0-14	0-9		0-21	0-14	0-9
0-1	1-2	1-2		0-1	0-1	0-1

MALE *FEMALE*

3.4   Limitations of All Three Procedures

Although the shuttle algorithm produced exact bounds for this example, the shuttle algorithm, and consequently the Fienberg ([6]) procedure and the F-bounds, are inexact, as follows. Examples 7b,c of [4] are 3-DIPTP exhibiting one or more non-integer continuous exact bounds. Because it is based on iterative improvement of integer upper bounds, in these situations the shuttle algorithm can come no closer than one unit larger than the exact integer bound, and therefore is incapable of achieving the exact integer bound.

The shuttle algorithm is not new: it was introduced by Schell [1] towards purported sufficient conditions on the 2-dimensional marginals for feasibility of 3-DPTP. By means of (16), Moravek-Vlach [2] show that the *Schell conditions* are necessary, but not sufficient, for the existence of a solution to the 3-DPTP. This counterexample is applicable here. Each 1-dimensional Fréchet lower bounds is negative, thus not effective. Each Bonferroni upper bound is too large to be sharp. No lower F-bound is positive. Iteration of the shuttle produces no improvement. Therefore, each procedure yields nominal lower (0) and upper (1) bounds for each entry. Each procedure converges. Consequently, all three procedures produce seemingly correct bounds when in fact no table exists. A simpler counterexample, (Example 2 of [5]), given by (17), appears in the next section.

$$A = \begin{pmatrix} 1 & 4 \\ 1 & 4 \\ 1 & 4 \\ 4 & 1 \\ 4 & 1 \\ 4 & 1 \\ 4 & 1 \\ 3 & 2 \end{pmatrix}, \quad B = \begin{pmatrix} 1 & 7 \\ 2 & 6 \\ 7 & 1 \\ 6 & 2 \\ 6 & 2 \end{pmatrix}, \quad C = \begin{pmatrix} 1 & 1 & 1 & 1 & 1 & 1 & 1 & 1 \\ 1 & 1 & 1 & 1 & 1 & 1 & 1 & 1 \\ 1 & 1 & 1 & 1 & 1 & 1 & 1 & 1 \\ 1 & 1 & 1 & 1 & 1 & 1 & 1 & 1 \\ 1 & 1 & 1 & 1 & 1 & 1 & 1 & 1 \end{pmatrix}. \quad (16)$$

4   The Roehrig and Chowdhury Network Models

Roehrig et al. (1999) [8] and Chowdhury et al. (1999) [9] offer network models for computing exact bounds on internal entries in 3-dimensional tables. Network models are extremely convenient and efficient, and most important enjoy the *integrality property*, viz., integer constraints (viz., 2-dimensional marginal totals) assure integer optima. Network models provide a natural mechanism and language in which to express 2-dimensional tables, but most generalizations beyond 2-dimensions are apt to fail.

Roehrig et al. [8] construct a network model for 2x2x2 tables and claim that it can be generalized to all 3-dimensional tables. This must be false. Cox ([4]) shows that the class of 3-dimensional tables (of size $d_1 x d_2 x d_3$) representable as a network is the set of tables for which $d_i < 3$ for at least one index i. This is true because, if all $d_i \geq 3$, it is possible to construct a 3-dimensional table with integer marginals whose corresponding polytope has non-integer vertices ([4]), which contradicts the integrality property.

Chowdhury et al. [9] address the following problem related to 3-DIPTP, also appearing in [8]. Suppose that the NSO releases only two of the three sets of 2-dimensional marginal totals (**A** and **B**), but not the third set (**C**) or the internal entries n_{ijk}. Is it possible to obtain exact lower and upper bounds for the remaining marginals (**C**)? The authors construct d_3 independent networks corresponding to the 2-dimensional tables defined by the appropriate (**A**, **B**) pairs and provide a procedure for obtaining exact bounds.

Unfortunately, this problem is quite simple and can be solved directly without recourse to networks or other mathematical formalities. In particular, the F-bounds of Section 2 suffice, as follows. Observe that, absent the **C**-constraints, the minimum (maximum) feasible value of a **C**-marginal total C_{ij} equals the sum of the minimum (maximum) values for the corresponding internal entries n_{ijk}. As the n_{ijk} are subject only to 2-dimensional constraints within their respective k-planes, then exact bounds for each n_{ijk} are precisely its 2-dimensional lower and upper F-bounds. These can be computed at sight and added along the k-dimension thus producing the corresponding **C**-bounds without recourse to networks or other formulations.

The Chowdhury et al. [9] method is insensitive to infeasibility, again demonstrated by example (16): all Fréchet lower and nominal upper bounds computed in the two 2-dimensional tables defined by k = 1 and k = 2 contain the corresponding $\mathbf{C}_{ij}$ (equal to 1 in all cases), but there is no underlying table as the problem is infeasible. Insensitivity is also demonstrated by Example 2 of [5], also presented at the 1998 conference, viz.,

$$
\mathbf{A} = \mathbf{B} = \mathbf{C} = \begin{pmatrix} 1 & 1 & 3 \\ 3 & 1 & 1 \\ 1 & 3 & 1 \end{pmatrix} .
\tag{17}
$$

5   Fractional Extremal Points

Theorem 4.3 of [4] demonstrates that the only multi-dimensional integer planar transportation problems (*m-DIPTP*) for which integer extremal points are assured are those of size 2^{m-2}xbxc. In these situations, linear programming methods can be relied upon to produce exact integer bounds on entries, and to do so in a computationally efficient manner even for problems of large dimension or size. The situation for other *integer problems of transportation type*, viz., m-dimensional contingency tables subject to k-dimensional marginal totals, k = 0, 1, ..., m-1, is quite different: until a direct connection can be demonstrated between exact continuous bounds on entries obtainable from linear programming and exact integer bounds on entries, linear programming will remain an unreliable tool for solving multi-dimensional problems of transportation type. Integer programming is not a viable option for large dimensions or size or repetitive use. Methods that exploit unique structure of certain subclasses of tables are then appealing, though possibly of limited applicability.

Dobra-Fienberg [10] present one such method, based on notions from mathematical statistics and graph theory. Given an m-dimensional integer problem of transportation type and specified marginal totals, if these marginals form a *set of sufficient statistics* for a specialized log-linear model known as a *decomposable graphical model*, then the model is feasible and exact integer bounds can be obtained from straightforward formulae. These formulae yield, in essence, the F- and Bonferroni bounds. The reader is referred to [10] for details, [11] for details on log-linear models, and [12] for development of graphical models.

The m-dimensional planar transportation problem considered here, $m > 2$, corresponds to the *no m-factor effect* log-linear model, which is not graphical, and consequently the Dobra-Fienberg method [10] does not apply to problems considered here.

The choice here and perhaps elsewhere of the 3- or m-DIPTP as the initial focus of study for bounding problems is motivated by the following observations. If for reasons of confidentiality the NSO cannot release the full m-dimensional tabulations (viz., the m-dimensional marginal totals), then its next-best strategy is to release the (m-1)-dimensional marginal totals, corresponding to the m-DIPTP. If it is not possible to release all of these marginals, perhaps the release strategy of Chowdhury et al. [9] should be investigated. Alternatively, release of the (m-2)-dimensional marginals might be considered. This strategy for release is based on the principle of releasing the most specific information possible without violating confidentiality. Dobra-Fienberg offers a different approach, a class of marginal totals, perhaps of varying dimension, that can be released while assuring confidentiality via easy computation of exact bounds on suppressed internal entries.

A formula driven bounding method is valuable for large problems and for repetitive, large scale use. Consider the m-dimensional integer problem of transportation type specified by its 1-dimensional marginal totals. In the statistics literature, this is known as the *complete independence* log-linear model [11]. This model, and in particular the 3-dimensional complete independence model, is graphical and decomposable. Thus, exact bounding can be achieved using Dobra-Fienberg.

Such problems can exhibit non-integer extremal points. For example, consider the 3x3x3 complete independence model with 1-dimensional marginal totals given by the vector:

$$\left(\sum_{i,j} n_{ijk}\right) = \left(\sum_{i,k} n_{ijk}\right) = \left(\sum_{j,k} n_{ijk}\right) = (2 \ 1 \ 2) \ . \tag{18}$$

Even though all continuous exact bounds on internal entries in (18) are integer, one extremal point at which n_{312} is maximized (at 1) contains four non-integer entries and another contains six. Bounding using linear programming would only demonstrate that $n_{312} = 1$ is the continuous, not the integer, maximum if either of these extremal points were exhibited. A strict network formulation is not possible because networks exhibit only integer extremal points (although use of networks with side constraints is under investigation). Integer programming is undesirable for large or repetitive applications. Direct methods such as Dobra-Fienberg may be required. A drawback of Dobra-Fienberg is that it applies only in specialized cases.

6   Discussion

In this paper we have examined the problem of determining exact integer bounds for entries in 3-dimensional integer planar transportation problems. This research was motivated by previous papers presenting heuristic approaches to similar problems that failed in some way to meet simple criteria for performance or reasonableness. We examined these and other approaches analytically and demonstrated one's superiority. We demonstrated that this method is imperfect and a reformulation of a method from the operations literature of the 1950s. We demonstrated that these methods are insensitive to infeasibility and can produce meaningless results otherwise undetected. We demonstrated that a method purported to generalize from 2x2x2 tables to all 3-dimensional tables could not possibly do so. We demonstrated that a problem posed and solved using networks in a *Management Science* paper can be solved by simple, direct means without recourse to mathematical programming. We illustrated the relationship between computing integer exact bounds, the presence of non-integer extremal points and the applicability of mathematical programming formulations such as networks.

NSOs must rely on automated statistical methods for operations including estimation, tabulation, quality assurance, imputation, rounding and disclosure limitation. Algorithms for these methods must converge to meaningful quantities. In particular, these procedures should not report meaningless, misleading results such as seemingly correct bounds on entries when no feasible values exist. These risks are multiplied in statistical data base settings where data from different sources often are combined. Methods examined here for bounds on suppressed internal entries in 3-dimensional contingency tables fail this requirement because they are heuristic and based on necessary, but not sufficient, conditions for the existence of a solution to the 3-DPTP. In addition, most of these methods fail to ensure exact bounds, and are incapable of identifying if and when they do in fact produce an exact bound. Nothing is gained by extending these methods to higher dimensions.

Disclaimer

Opinions expressed are solely those of the author and are not intended to represent policy or practices of the National Center for Health Statistics, Centers for Disease Control and Prevention, or any other organization.

References

1. Schell, E. Distribution of a product over several properties. Proceedings, 2nd Symposium on Linear Programming. Washington, DC (1955) 615-642
2. Moravek, J. and Vlach, M. (1967). On necessary conditions for the existence of the solution to the multi-index transportation problem. Operations Research 15, 542-545
3. Vlach, M. Conditions for the existence of solutions of the three-dimensional planar transportation problem. Discrete Applied Mathematics 13 (1986) 61-78
4. Cox, L. (2000). On properties of multi-dimensional statistical tables. Manuscript (April 2000) 29 pp.

5. Cox, L. Invited Talk: Some remarks on research directions in statistical data protection. Statistical Data Protection: Proceedings of the Conference.EUROSTAT, Luxemburg (1999)163-176

6. Fienberg, S. Fréchet and Bonferroni bounds for multi-way tables of counts with applications to disclosure limitation. Statistical Data Protection: Proceedings of the Conference. EUROSTAT, Luxembourg (1999) 115-129

7. Buzzigoli, L. and Giusti, A. An algorithm to calculate the lower and upper bounds of the elements of an array given its marginals. Statistical Data Protection: Proceedings of the Conference. EUROSTAT, Luxemburg (1999) 131-147

8. Roehrig, S., Padman, R., Duncan, G., and Krishman, R. Disclosure detection in multiple linked categorical datafiles: A unified network approach. Statistical Data Protection: Proceedings of the Conference. EUROSTAT, Luxembourg (1999) 149-162

9. Chowdhury, S., Duncan, G., Krishnan, R., Roehrig, S., and Mukherjee, S. Disclosure detection in multivariate categorical databases: Auditing confidentiality protection through two new matrix operators. Management Science 45 (1999) 1710-1723

10. Dobra, A. and S. Fienber, S.. Bounds for cell entries in contingency table given marginal totals and decomposable graphs. Proceedings of theNational Academy of Sciences 97 (2000) 11185-11192

11. Bishop, Y., Fienberg, S., and Holland, P. Discrete Multivariate Analysis: Theory and Practice. Cambridge, MA: M.I.T. Press (1975)

12. Lauritzen, S. Graphical Models. Oxford: Clarendon Press (1996)

Extending Cell Suppression to Protect Tabular Data against Several Attackers*

Juan José Salazar González

DEIOC, Faculty of Mathematics, University of La Laguna
Av. Astrofísico Francisco Sánchez, s/n; 38271 La Laguna, Tenerife, Spain
Tel: +34 922 318184; Fax: +34 922 318170
jjsalaza@ull.es

Abstract. This paper presents three mathematical models for the problem of finding a cell suppression pattern minimizing the loss of information while guaranteeing protection level requirements for different sensitive cells and different intruders. This problem covers a very general setting in Statistical Disclosure Control, and it contains as particular cases several important problems like, e.g., the so-called "common respondent problem" mentioned in Jewett [9]. Hence, the three models also applies to the common respondent problem, among others. The first model corresponds to bi-level Mathematical Programming. The second model belongs to Integer Linear Programming (ILP) and could be used on small-size tables where some nominal values are known to assume discrete values. The third model is also an ILP model valid when the nominal values of the table are continuous numbers, and with the good advantage of containing an small number of variables (one 0-1 variable for each cell in the table). On the other hand, this model has a bigger number of linear inequalities (related with the number of sensitive cells and the number of attackers). Nevertheless, this paper addresses this disadvantage which is overcame by a dynamic generation of the important inequalities when necessary. The overall algorithm follows a modern Operational Research technique known as branch-and-cut approach, and allows to find optimal solutions to medium-size tables. On large-size tables the approach can be used to find near-optimal solutions. The paper illustrates the procedure on an introductory instance.
The paper ends pointing another alternative methodology (closely related to the one in Jewett [9]) to produce patterns by shrinking all the different intruders into a single one, and compares it with the classical single-attacker methodology and with the above multi-attacker methodology.

1 Introduction

Cell suppression is one of the most popular techniques for protecting sensitive information in statistical tables, and it is typically applied to 2- and 3-dimensional

---

* Work supported by the European project IST-2000-25069, "Computational Aspects of Statistical Confidentiality" (CASC).

J. Domingo-Ferrer (Ed.): Inference Control in Statistical Databases, LNCS 2316, pp. 34–58, 2002.

tables whose entries (*cells*) are subject to marginal totals. The standard cell suppression technique is based on the idea of protecting the sensitive information by hiding the values of some cells with a symbol (e.g. *). The aim is that a set of potential intruders could not guess (exactly or approximately) any one of the hiding values by only using the published values and some a-priori information. The only assumption of this work is that this a-priori information must be formulated as a linear system of equations or inequations with integer and/or continuous variables. For example, we allow the intruder to know bounds on the hiding values (as it happens with bigger contributors to each cell) but not probability distributions on them. Notice that different intruders could know different a-priori information.

The aim is considered so complex that there are in literature only heuristic approaches (i.e., procedures providing approximated —probably overprotected— suppression patterns) for special situations. For example, a relevant situation occurs when there is an entity which contributes to several cells, leading to the so-called *common respondent problem*. Possible simplifications valid for this situation consist on replacing all the different intruders by one stronger attacker with "protection capacities" associated to the potential secondary suppressions (see, e.g., Jewett [9] or Sande [17] for details), or on aggregating some sensitive cells into new "union" cells with stronger protection level requirements (see, e.g., Robertson [15]).

This paper presents the first mathematical models for the problem in the general situation (i.e, without any simplification) and a first exact algorithm for the resolution. The here-proposed approach looks for a suppression pattern with minimum loss of information and which guarantees all the protection requirements against all the attackers. It is also a practical method to find an optimal solution using modern tools from Mathematical Programming, a well-established methodology (see, e.g., [13]). The models and the algorithm apply to a very general problem, containing the common respondent problem as a particular case. Section 2 illustrates the classical cell suppression methodology by means of a (very simple) introductory example, and Section 3 describes the more general multi-attacker cell suppression methodology. Three mathematical models are described in Section 4, the last one with one decision variable for each cell and a large number of constraints. Section 5 proposes an algorithm for finding an optimal solution of the model using the third model, and Section 6 illustrates how it works on the introductory example. Section 7 presents a relaxed methodology based on considering one worse-case attacker with the information of all the original intruders, leading to an intermediate scheme that could be applied with smaller computational effort. This scheme considers the "protection capacities" in Jewett [9] and provides overprotected patterns. Finally, Section 8 compares the classical, the multi-attacker and the intermediate methodologies, and Section 9 summarizes the main ideas of this paper and point out a further extension of the multi-attacker models.

2   Classical Cell Suppression Methodology

A common hypothesis in the classical cell suppression methodology (see, e.g., Willenborg and De Waal [18]) is that there is *only one* attacker interested in the disclosure of all sensitive cells. We next introduce the main concepts of the methodology through a simple example.

	A	B	C	Total
Activity I	20	50	10	80
Activity II	8	19	**22**	49
Activity III	17	32	12	61
Total	45	101	44	190

Fig. 1. Investment of enterprises by activity and region

	A	B	C	Total
Activity I	20	50	10	80
Activity II	*	19	*	49
Activity III	*	32	*	61
Total	45	101	44	190

Fig. 2. A possible suppression pattern

Figure 1 exhibits a statistical table giving the investment of enterprises (per millions of guilders), classified by activity and region. For simplicity, the cell corresponding to Activity i (for each $i \in \{I, II, III\}$) and Region j (for each $j \in \{A, B, C\}$) will be represented by the pair (i, j). Let us assume that the information in cell (II, C) is confidential, hence it is viewed as a *sensitive cell* to be suppressed. By using the marginal totals the *attacker* can however recompute the missing value in the sensitive cell, hence other table entries must be suppressed as well, e.g., those of Figure 2. With this choice, the attacker cannot disclosure the value of the sensitive cell exactly, although he/she can still compute a range for the values of this cell which are consistent with the published entries. Indeed, from Figure 2 the minimum value $y_{II,C}^-$ for the sensitive cell (II, C) can be computed by solving a Linear Programming (LP) model in which the values $y_{i,j}$ for the suppressed cells (i, j) are treated as unknowns, namely

$$y_{II,C}^- := \min y_{II,C}$$

subject to

$$
\begin{aligned}
y_{II,A} \quad\quad\quad +y_{II,C} \quad\quad\quad &= 30 \\
y_{III,A} \quad\quad\quad +y_{III,C} &= 29 \\
y_{II,A} +y_{III,A} \quad\quad\quad\quad\quad &= 25 \\
y_{II,C} +y_{III,C} &= 34
\end{aligned}
$$

$$y_{II,A} \geq 0 \,,\; y_{III,A} \geq 0 \,,\; y_{II,C} \geq 0 \,,\; y_{III,C} \geq 0.$$

Notice that the right-hand side values are known to the attacker, as they can be obtained as the difference between the marginal and the published values in a row/column. We are also assuming that the attacker knows that a missing value is non-negative, i.e., 0 and infinity are known "external bounds" for suppressions.

The maximum value $y_{II,C}^{+}$ for the sensitive cell can be computed in a perfectly analogous way, by solving the linear program of maximizing $y_{II,C}$ subject to the same constraints as before. Notice that each solution of this common set of constraints is a congruent table according with the published suppression pattern in Figure 2 and with the extra knowledge of the external bounds (non-negativity on this example).

In the example, $y_{II,C}^{-} = 5$ and $y_{II,C}^{+} = 30$, i.e., the sensitive information is "protected" within the *protection interval* [5,30]. If this interval is considered sufficiently wide by the statistical office, then the sensitive cell is called *protected*; otherwise Figure 2 is not a valid suppression pattern and new complementary suppressions are needed.

Notice that the extreme values of the computed interval [5, 30] are only attained if the cell (II, A) takes the quite unreasonable values of 0 and 25. Otherwise, if the external bounds for each suppressed cell are assumed to be ±50% of the nominal value, then the solution of the new two linear programs results in the more realistic protection interval [18, 26] for the sensitive cell. That is why it is very important to consider good estimations of the external bounds known for the attacker on each suppressed cell when checking if a suppression pattern protects (or not) each sensitive cell of the table. As already stated, in the above example we are assuming that the external bounds are 0 and infinity, i.e., the only knowledge of the attacker on the unknown variables is that they are non-negative numbers.

To classify the computed interval $[y_p^{-}, y_p^{+}]$ around a nominal value a_p of a sensitive cell p as "sufficiently wide" or not, the statistical office must provide us with three parameters for each sensitive cell:

- an *upper protection level* representing the minimum allowed value to $y_p^{+} - a_p$;
- a *lower protection level* representing the minimum allowed value to $a_p - y_p^{-}$;
- an *sliding protection level* representing the minimum allowed value to $y_p^{+} - y_p^{-}$.

For example, if 7, 5 and 0 are the upper, lower and sliding protection levels, respectively, then the interval [5, 30] is "sufficiently wide", and therefore pattern in Figure 2 is a valid solution for the statistical office (assuming the external bounds are 0 and infinity).

The statistical office then aims at finding a valid suppression pattern protecting all the sensitive cells against the attacker, and such that the loss of information associated with the suppressed entries is minimized. This results into a combinatorial optimization problem known as the (classical) *Cell Suppression Problem*, or CSP for short. CSP belongs to the class of the strongly $\mathcal{NP}$-hard problems (see, e.g., Kelly et al. [12], Geurts [7], Kao [10]), meaning that it is very unlikely that an algorithm for the exact solution of CSP exists, which guarantees an efficient (i.e., polynomial-time) performance for all possible input instances.

Previous works on the classical CSP from the literature mainly concentrate on 2-dimensional tables with marginals. Heuristic solution procedures have been proposed by several authors, including Cox [1,2], Sande [16], Kelly et al. [12], and Carvalho et al. [3]. Kelly [11] proposed a mixed-integer linear programming formulation involving a huge number of variables and constraints (for instance, the formulation involves more than 20,000,000 variables and 30,000,000 constraints for a two-dimensional table with 100 rows, 100 columns and 5% sensitive entries). Geurts [7] refined this model, and reported computational experiences on small-size instances, the largest instance solved to optimality being a table with 20 rows, 6 columns and 17 sensitive cells. (the computing time is not reported; for smaller instances, the code required several thousand CPU seconds on a SUN Spark 1+ workstation). Gusfield [8] gave a polynomial algorithm for a special case of the problem. Heuristics for 3-dimensional tables have been proposed in Robertson [14], Sande [16], and Dellaert and Luijten [4]. Very recently, Fischetti and Salazar [5] proposed a new method capable of solving to proven optimality, on a personal computer, 2-dimensional tables with about 250,000 cells and 10,000 sensitive entries. An extension of this methodology capable of solving to proven optimality real-world 3- and 4-dimensional tables is presented in Fischetti and Salazar [6].

3   Multi-attacker Cell Suppression Methodology

The classical cell suppression methodology has several disadvantages. One of them concerns with the popular hypothesis that the table must be protected against *one* attacker. To be more precise, with the above assumption the attacker is supposed to be one external intruder with no other information different than the structure of the table, the published values and the external bounds on the suppressed values. In practice, however, there are also other types of attackers, like some special respondents (e.g., different entities contributing to different cell values). We will refer to those ones as *internal attackers*, while the above intruder will be refereed as *external attacker*. For each internal attacker there is an additional information concerning his/her own contribution to the table. To be more precise, in the above example, if cell (II, A) has only one respondent, the output from Figure 2 could be protected for an external attacker but not from this potential internal attacker. Indeed, the respondent contributing to cell (II, A) knows that $y_{II,A} \geq 8$ (and even that $y_{II,A} = 8$ if he/she also knows that he/she is the only contributor to such cell). This will allow him/her to compute a more narrow protection interval for the sensitive cell (II, C) from Figure 2, even if it is protected for the external attacker.

In order to avoid this important disadvantage of the classical cell suppression methodology, this paper proposes an extension of the mathematical model presented in Fischetti and Salazar [6]. The extension determines a suppression pattern protected against external and internal attackers and it will be described in the next section.

The classical Cell Suppression will be also referred through this article as *Single-attacker CSP*, while the new proposal will be named as *Multi-attacker CSP*. The hypothesis of the new methodology is that we must be given, not only with the basic information (nominal values, loss of information, etc), but also with a set of attackers K and for each one the specific information he/she has (i.e., his/her own bounds on unpublished values).

Notice that if K contains only the external attacker, then the multi-attacker CSP reduces to the single-attacker CSP. Otherwise it could happens that some attackers could be not considered since a suppression pattern that protect sensitive information against one attacker could also protect the table against some others. This is, for example, the case when there are two attackers with the same protection requirements, but one knows tighter bounds; then it is enough to protect the table against the stronger attacker. For each attacker, a similar situation occurs with the sensitive cells since protecting some of them could imply to protect others. Therefore, a clever *preprocessing* is always required to reduce as much as possible the number of protection level requirements and the number of attackers.

The availability of a well-implemented preprocessing could help also in the task of setting the potential internal attackers. Indeed, notice that in literature there have been developed several rules to establish the sensitive cells in a table (e.g., the dominance rule) but the same effort does not exist to establish the attackers. Within the preprocessing a proposal could be to consider each respondent in a table as a potential intruder, and then simply apply the preprocessing to remove the dominated ones. In theory this approach could lead to a huge number of attackers, but in practice it is expected a number of attackers not bigger than the number of cells in the table (and hopefully much smaller).

Another important observation is the following. Considering a particular sensitive cell, the statistical office could also be interested on providing different protection levels for each attacker. For example, suppose that the statistical office requires a lower protection level of 15 and an upper protection level of 5 for the sensitive cell (II, C) in the introductory example (Figure 1) against an external attacker. If the sensitive cell is the sum of the contribution from two respondents, one providing 18 units and the other providing 4 units, the statistical office could be also interested on requiring an special upper protection level of 20 against the biggest contributor to the sensitive cell (because he/she is a potential attacker with the extra knowledge that the sensitive cell contains at least value 18). Indeed notice that it does not make sense to protect the sensitive cell against the internal attacker within a lower protection requirement of at least 5, since he/she already knows a lower bound of 18. In other words, it makes sense that the statistical office wants different protection levels for different attackers, with the important assumption that each protection level must be smaller than the correspondent bound. The following section describes Mathematical Programming models capturing all these features.

4   Mathematical Models

Let us consider a table $[a_i, i \in I]$ with $n := |I|$ cells. It can be a k-dimensional, hierarchical or linked table. Since there are marginals, the cells are linked through some equations indexes by J, and let $[\sum_{i \in I} m_{ij} y_i = b_j, j \in J]$ the linear system defined by such equations. (Typically $b_j = 0$ and $m_{ij} \in \{-1, 0, +1\}$ with one -1 in each equation.) We are also given a weight w_i for each cell $i \in I$ for the loss of information incurred if such cell is suppressed in the final suppression pattern. Let $P \subset I$ the set of sensitive cells (and hence the set of primary suppression). Finally, let us consider a set K of potential attackers. Associated to each attacker $k \in K$, we are given with the external bounds (lb_i^k, ub_i^k) known by the attacker on each suppressed cell $i \in I$, and with the three protection levels $(upl^{kp}, lpl^{kp}, spl^{kp})$ that the statistical office requires to protect each sensitive cell $p \in P$ against such attacker k. From the last observation in the previous section, we will assume

$$lb_p^k \leq a_p - lpl^{kp} \leq a_p \leq a_p + upl^{kp} \leq ub_p^k$$

and

$$ub_p^k - lb_p^k \geq spl^{kp}$$

for each attacker k and each sensitive cell p.

Then, the optimization problem associated to the Cell Suppression Methodology can be modeled as follows. Let us consider a binary variable x_i associated to each cell $i \in I$, assuming value 1 if such cell must be suppressed in the final pattern, or 0 otherwise. Notice that the attacker will minimize and maximize unknown values on the set of consistent tables, defined by:

$$\sum_{i \in I} m_{ij} y_i = b_j \quad , j \in J$$
$$lb_i^k \leq y_i \leq ub_i^k \quad , i \in I \text{ when } x_i = 1$$
$$y_i = a_i \qquad , i \in I \text{ when } x_i = 0,$$

equivalently represented as the solution set of the following linear system:

$$\left. \begin{array}{r} \sum_{i \in I} m_{ij} y_i = b_j \qquad\qquad , j \in J \\ a_i - (a_i - lb_i^k) x_i \leq y_i \leq a_i + (ub_i^k - a_i) x_i \, , i \in I. \end{array} \right\} \quad (1)$$

Therefore, our optimization problem is to find a value for each x_i such that the total loss of the information in the final pattern is minimized, i.e.:

$$\min \sum_{i \in I} w_i x_i \quad (2)$$

subject to, for each sensitive cell $p \in P$ and for each attacker $k \in K$,

– the upper protection requirement must be satisfied, i.e.:

$$\max \{y_p : (1) \text{ holds } \} \geq a_p + upl^{kp} \quad (3)$$

– the lower protection requirement must be satisfied, i.e.:

$$\min \{y_p : (1) \text{ holds } \} \leq a_p - lpl^{kp} \qquad (4)$$

– the sliding protection requirement must be satisfied, i.e.:

$$\max \{y_p : (1) \text{ holds } \} - \min \{y_p : (1) \text{ holds } \} \geq spl^{kp} \qquad (5)$$

Finally, each variable must assume value 0 or 1, i.e.:

$$x_i \in \{0, 1\} \qquad \text{for all } i \in I. \qquad (6)$$

Mathematical model (2)–(6) contains all the requirements of the statistical office (according with the definition given in Section 1), and therefore a solution $[x_i^*, i \in I]$ defines an optimal suppression pattern. The inconvenient is that it is not an easy model to be solved, since it does not belong to the standard Mixed Integer Linear Mathematical Programming. In fact, the existence of optimization problems as constraints of a main optimization problem classifies the model in the so-called "Bilevel Mathematical Programming", which does not contain efficient algorithms to solve model (2)–(6) even of small sizes. Observe that the inconvenience of model (2)–(6) is not the number of variables (which is at most the number of cells both for the master optimization problem and for each subproblem in the second level), but the fact there are nested optimization problems in two levels. The better way to avoid the direct resolution it is by looking for a transformation into a classical model in Integer Programming.

A first idea arises by observing that the optimization problem in condition (3) can be replaced by the existence of a table $[f_i^{kp}, i \in I]$ such that it is congruent (i.e., it satisfies (1)) and it guarantees the upper protection level requirement, i.e.:

$$f_p^{kp} \geq a_p + upl^{kp}.$$

In the same way, the optimization problem in condition (4) can be replaced by the existence of a table $[g_i^{kp}, i \in I]$ such that it is also congruent (i.e., it satisfies (1)) and it guarantees the lower protection level requirement, i.e.:

$$g_p^{kp} \leq a_p - lpl^{kp}.$$

Finally, the two optimization problems in condition (5) can be replaced by the above congruent tables if they guarantee the sliding protection level, i.e.:

$$f_p^{kp} - g_p^{kp} \geq spl^{kp}.$$

Figure 3 shows a first attempt to have an integer linear model.

Clearly, this new model is a Mixed Integer Linear Programming model, and therefore —in theory— there are efficient approaches to solve it. Nevertheless, the number of new variables (f_i^{kp} and g_i^{kp}) is really huge even on small tables. For example, the model associated with a table with 100×100 cells with 1%

$$\min \sum_{i \in I} w_i x_i$$

subject to:

$$\sum_{i \in I} m_{ij} f_i^{kp} = b_j \qquad\qquad \text{for all } j \in J$$

$$a_i - (a_i - lb_i^k)x_i \le f_i^{kp} \le a_i + (ub_i^k - a_i)x_i \quad \text{for all } i \in I$$

$$\sum_{i \in I} m_{ij} g_i^{kp} = b_j \qquad\qquad \text{for all } j \in J$$

$$a_i - (a_i - lb_i^k)x_i \le g_i^{kp} \le a_i + (ub_i^k - a_i)x_i \quad \text{for all } i \in I$$

$$f_p^{kp} \ge a_p + upl^{kp}$$

$$g_p^{kp} \le a_p - lpl^{kp}$$

$$f_p^{kp} - g_p^{kp} \ge spl^{kp}$$

for all $p \in P$ and all $k \in K$, and also subject to:

$$x_i \in \{0, 1\} \qquad\qquad \text{for all } i \in I.$$

Fig. 3. First ILP model for multi-attacker CSP

sensitive and 100 attackers would have millions of variables. Therefore, it is necessary another approach to transform model (2)–(6) without adding so many additional variables.

An alternative approach which does not add any additional variable follows the idea described in Fischetti and Salazar [6] for the classical cell suppression problem (i.e., with one attacker). Based on the Farkas' Lemma, it is possible to replace the second level problems of model (2)–(6) by linear constraints on the x_i variables. Indeed, assuming that values y_i in a congruent table are continuous numbers, the two linear programs in conditions (3)–(5) can be rewritten in their dual format. More precisely, by Dual Theory in Linear Programming

$$\max \{y_p : (1) \text{ holds } \}$$

is equivalent to

$$\min \sum_{j \in J} \gamma_j b_j + \sum_{i \in I} [\alpha_i(a_i + (ub_i^k - a_i)x_i) - \beta_i(a_i - (a_i - lb_i^k)x_i)]$$

subject to

$$
\left.\begin{array}{r}
\alpha_p - \beta_p + \sum_{j \in J} m_{pj}\gamma_j = 1 \\
\alpha_i - \beta_i + \sum_{j \in J} m_{ij}\gamma_j = 0 \quad \text{for all } i \in I \setminus \{p\} \\
\alpha_i \geq 0 \quad \text{for all } i \in I \\
\beta_i \geq 0 \quad \text{for all } i \in I \\
\gamma_j \text{ unrestricted in sign} \quad \text{for all } j \in J.
\end{array}\right\} \tag{7}
$$

Because of (7) and $[a_i, i \in I]$ is a consistent table, we have

$$
\sum_{j \in J} \gamma_j b_j + \sum_{i \in I}(\alpha_i a_i - \beta_i a_i) = \sum_{i \in I}\sum_{j \in J}\gamma_j m_{ij} a_i + \sum_{i \in I}(\alpha_i - \beta_i)a_i = a_p.
$$

Hence the above linear program can be rewritten as

$$
a_p + \min \sum_{i \in I}(\alpha_i(ub_i^k - a_i) + \beta_i(a_i - lb_i^k))x_i
$$

subject to $\alpha_i, \beta_i, \gamma_j$ satisfying (7).

From this observation, condition (3) can be now written as:

$$
\sum_{i \in I}(\alpha_i(ub_i^k - a_i) + \beta_i(a_i - lb_i^k))x_i \geq upl^{kp} \qquad \text{for all } \alpha_i, \beta_i, \gamma_j \text{ satisfying (7).}
$$

In other words, the last system defines a family of linear constraints, in the x-variables only, representing the condition (3) which concerns with the upper protection level requirement for sensitive cell p and attacker k.

Notice that this family contains in principle an infinite number of constraints, each associated with a different point $[\alpha_i, i \in I; \beta_i, i \in I; \gamma_j, j \in J]$ of the polyhedron defined by (7). However, it is well known that only the extreme points (and rays) of such polyhedron can lead to undominated constraints, i.e., a finite number of such constraints is sufficient to impose the upper protection level requirement for a given sensitive cell p and a give attacker k.

In a similar way, the optimization problem in (4) is:

$$
- \max\{-y_p : (1) \text{ holds }\},
$$

which, by Duality Theory, is equivalent to

$$
- \min \sum_{j \in J} \gamma_j b_j + \sum_{i \in I}[\alpha_i(a_i + (ub_i^k - a_i)x_i) - \beta_i(a_i - (a_i - lb_i^k)x_i)]
$$

subject to

$$
\left.\begin{array}{r}
\alpha_p - \beta_p + \sum_{j \in J} m_{pj}\gamma_j = -1 \\
\alpha_i - \beta_i + \sum_{j \in J} m_{ij}\gamma_j = 0 \quad \text{for all } i \in I \setminus \{p\} \\
\alpha_i \geq 0 \quad \text{for all } i \in I \\
\beta_i \geq 0 \quad \text{for all } i \in I \\
\gamma_j \text{ unrestricted in sign} \quad \text{for all } j \in J.
\end{array}\right\} \tag{8}
$$

Because of (8) and $[a_i, i \in I]$ is a consistent table, we have

$$\sum_{j \in J} \gamma_j b_j + \sum_{i \in I} (\alpha_i a_i - \beta_i a_i) = \sum_{i \in I} \sum_{j \in J} \gamma_j m_{ij} a_i + \sum_{i \in I} (\alpha_i - \beta_i) a_i = a_p.$$

Hence the above linear program can be rewritten as

$$-a_p - \min \sum_{i \in I} (\alpha_i (ub_i^k - a_i) + \beta_i (a_i - lb_i^k)) x_i$$

subject to $\alpha_i, \beta_i, \gamma_j$ satisfying (8).

From this observation, condition (4) can be now written as:

$$\sum_{i \in I} (\alpha_i (ub_i^k - a_i) + \beta_i (a_i - lb_i^k)) x_i \geq lpl^{kp} \qquad \text{for all } \alpha_i, \beta_i, \gamma_j \text{ satisfying (8).}$$

In other words, the last system defines a family of linear constraints, in the x-variables only, representing the condition (4) which concerns with the lower protection level requirement for sensitive cell p and attacker k.

As to the sliding protection level for sensitive cell i_k, the requirement is that

$$spl^{kp} \leq \max\{y_p : (1) \text{ hold }\} + \max\{-y_p : (1) \text{ hold }\}.$$

Again, by LP duality, this condition is equivalent to

$$spl^{kp} \leq$$

$$\min \left\{ \sum_{j \in J} \gamma_j b_j + \sum_{i \in I} [\alpha_i (a_i + (ub_i^k - a_i) x_i) - \beta_i (a_i - (a_i - lb_i^k) x_i)] : (7) \text{ holds } \right\} +$$

$$\min \left\{ \sum_{j \in J} \gamma_j b_j + \sum_{i \in I} [\alpha_i (a_i + (ub_i^k - a_i) x_i) - \beta_i (a_i - (a_i - lb_i^k) x_i)] : (8) \text{ holds } \right\}.$$

Therefore, the feasibility condition can now be formulated by requiring

$$spl^{kp} \leq \sum_{j \in J} (\gamma_j + \gamma_j') b_j +$$

$$\sum_{i \in I} [(\alpha_i + \alpha_i')(a_i + (ub_i - a_i) x_i) - (\beta_i + \beta_i')(a_i - (a_i - lb_i) x_i)]$$

$$\text{for all } \alpha, \beta, \gamma \text{ satisfying (7) and for all } \alpha', \beta', \gamma' \text{ satisfying (8),}$$

or, equivalently,

$$\sum_{i \in I} [(\alpha_i + \alpha_i')(ub_i - a_i) + (\beta_i + \beta_i')(a_i - lb_i)] x_i \geq spl^{kp}$$

$$\text{for all } \alpha, \beta, \gamma \text{ satisfying (7) and for all } \alpha', \beta', \gamma' \text{ satisfying (8).}$$

In conclusion, Figure 4 summarizes an alternative model to (2)–(6) with only the 0-1 variables. This model is not only a theoretical result, but it can be also used in practice to implement the effective algorithm described in the next section.

$$\min \sum_{i \in I} w_i x_i$$

subject to:

$$\sum_{i \in I} [\alpha_i (ub_i^k - a_i) + \beta_i (a_i - lb_i^k)] x_i \geq upl^{kp}$$

for all α, β, γ satisfying (7)

$$\sum_{i \in I} [\alpha_i' (ub_i^k - a_i) + \beta_i' (a_i - lb_i^k)] x_i \geq lpl^{kp}$$

for all α', β', γ' satisfying (8)

$$\sum_{i \in I} [(\alpha_i + \alpha_i')(ub_i - a_i) + (\beta_i + \beta_i')(a_i - lb_i)] x_i \geq spl^{kp}$$

for all α, β, γ satisfying (7) and
for all α', β', γ' satisfying (8),

for all $p \in P$ and all $k \in K$, and also subject to:

$$x_i \in \{0, 1\} \qquad \text{for all } i \in I.$$

Fig. 4. Second ILP model for multi-attacker CSP

5   An Algorithm for the Second ILP Model

Let us forget in a first step the integrability on the x_i-variables in order to solve the linear relaxation of the model in Figure 4. In a second step the integrability can be achieved through classical Mathematical Programming approaches (i.e., branch-and-bound, cutting-plane, branch-and-cut, etc.).

For the first step, even if the number of constraints in the model is finite, it is huge since it is order of the number of extreme points of polyhedra defined by (7) and (8). Hence, some special mechanisms are required to implicitly manage the amount without having all of them explicitly. In other words, it is required a procedure to generate on-the-flight the important constraints whenever they are necessary. This problem is known as *separation problem* and could be precisely defined as follows:

Given a solution $[x_i^*, i \in I]$ satisfying *some* linear constraints of model in Figure 4, does it satisfy *all* the linear constraints? If not, find at least one *violated* linear constraint.

It is known in Mathematical Programming (see, e.g., Nemhauser and Wolsey [13]) that the complexity of that separation problem is equivalent to the optimization problem (i.e., the problem of solving the linear relaxation with all the linear constraints). Therefore, out first objective is to solve the separation problem.

Whenever we have a (possibly fractional) solution $[x_i^*, i \in I]$ it is possible to check if all the linear constraints are satisfied by just solving two linear programming problems for each sensitive cell p and for each attacker k. In fact, by solving

$$\min \{y_p : (1) \text{ holds }\} \qquad \text{and} \qquad \max \{y_p : (1) \text{ holds }\}$$

with $x_i := x_i^*$ for all $i \in I$, it is easy to check if the three protection levels are (or not) satisfied. If they are, then $[x_i^*, i \in I]$ satisfied all the linear conditions (and hence all the protection level requirements even if it could be an unacceptable solution when it contains a non-integer component). Otherwise, the violated protection level provides a violated linear constraint defined by the dual solutions of the linear programs.

In conclusion, an algorithm to solve the linear relaxation of model in Figure 4 consists on an iterative procedure. At each iteration, two linear programs are solved for each sensitive cell and for each attacker. If all the protection levels are satisfied for all the sensitive cells and for all the attackers, then the algorithm stops. Otherwise, the dual variables of the linear programs non satisfying a protection level define a violated constraint. This constraint is added to the current linear program and a next iteration follows.

After we obtain a (possible fractional) solution $[x_i^*, i \in I]$ satisfying all the linear constraints (i.e., an optimal solution of the complete linear relaxation) then a further mechanism is needed to enforce integrability on the decision variables x_i. A possibility is to follows a branching procedure by fixing a non-integer component to 0 in one branch and to 1 in another branch. This leads to exploring a binary tree of subproblems. A better possibility is to continue improving the lower bound before branching, hence following a sol-called *branch-and-bound* scheme (see, e.g., Nemhauser and Wolsey [13]). To to that, we need additional valid linear constraints. A new family of constraints to strengthen a linear relaxation consists on observing that even non-sensitive cells must be protected if they are suppressed. In other words, since the loss of information of a cell is positive, if it is suppressed then the precise value must be protected. This can be imposed on a cell l by requiring a sliding protection level of ϵx_l for a small number $\epsilon > 0$. Notice that the sliding protection level is active only when x_l is also positive. Then, the linear constraints for imposing these additional requirements are

$$\sum_{i \in I} [(\alpha_i + \alpha_i')(ub_i - a_i) + (\beta_i + \beta_i')(a_i - lb_i)] x_i \geq \epsilon x_l$$

for all (α, β, γ) satisfying (7) and for all $(\alpha', \beta', \gamma')$ satisfying (8) where $p := l$ and all attackers k must be considered. Because each variable must be 0 or 1, it is possible to round down the right-hand-side coefficients, so this new family of constraints can be written as follows:

$$\sum_{i \in I : (\alpha_i + \alpha'_i)(ub_i - a_i) + (\beta_i + \beta'_i)(a_i - lb_i) \neq 0} x_i \geq x_l \tag{9}$$

for all (α, β, γ) satisfying (7) and for all $(\alpha', \beta', \gamma')$ satisfying (8), where $p := l$ and all attackers k must be considered. Given a solution $[x_i^*, i \in I]$, an approach to solve the separation problem of this new family of constraints consists on solving the two linear programs

$$\min \{y_l : (1) \text{ holds }\} \qquad \text{and} \qquad \max \{y_l : (1) \text{ holds }\}$$

for each nonsensitive cell l with $x_l^* > 0$ and for each attacker k. If the difference between the objective values is zero then a violated inequality in (9) is found by considering the dual solutions for the solved linear programs. Therefore, once again, the huge number of constraints can be easily managed within a cutting-plane algorithm.

In any case, the original linear constraints (even extended with the new inequalities) are not enough for guarantee integrality on the decision variables x_i, so in any case the branching phase could be required to obtain an integer solution. In practice, however, this is not always the case and several instances can be protected just by using the two considered families of constraints. This happens when protecting the introductory example, as Section 6 shows.

The overall algorithm is called branch-and-cut algorithm, and there are several additional tricks to speed up the performance. For example, if possible, it is better to find more than one violated constraints before going to the next iteration. It is helpful to start the initial iteration with some constrains that are suspected to be necessary (as the ones defined from the rows and columns where the sensitive cells are). All these possibly important inequalities are saved in the so-called *constraint pool*.

As described, the algorithm provides a lower bound on the loss of information by solving a linear relaxation. It is also helpful to have heuristic approaches for generating integer valid solutions at the beginning and during the branch-tree enumeration, even if they are overprotected. In fact, this approximated solutions produce an upper bound on the loss of information, and allow to do not explore some branches of the branch-and-bound scheme. Moreover, they provide an stopping criterion since the overall algorithm finishes when both the lower and the upper bounds coincide. Finally, even if the algorithm is interrupted before completion, an approximated solution is provided to the statistical office within a worse-case measure of the distance to an optimal solution (the gap between the lower and the upper bounds).

Therefore, the proposed approach always guarantees a feasible suppression pattern, which satisfies all the protection level requirements on all sensitive cells against all attackers. If the approach is interrupted before completion, it provides a heuristic solution, while if it is finished with identical upper and lower bounds, it provides an optimal solution.

6   An Example for the Second ILP Model

We will next illustrate how the algorithm described in Section 5 proceeds on the simple instance presented in the introduction. For simplicity, we will assume to have only one sensitive cell and one attacker. A more complex instance (e.g., with more than one sensitive cell and/or more than one attacker) could be also solved by just repeating the here-illustrated procedure for each pair of sensitive cell and attacker.

Let us suppose that all the 15 non-sensitive cells are allowed to be potential secondary suppressions in a pattern protecting the sensitive cell, and suppose that the loss of information is defined as equal to the nominal value of a cell (i.e., $w_i = a_i$ for all i). Each cell will be represented with a two-digit number, the first one corresponding to the row and the second corresponding to the column. For example, cell with index $i = 23$ represents the cell in row 2 (activity II) and column 3 (region C). Let us consider an attacker known external bounds $lb_i = 0$ and $ub_i = 1000$ on each unpublished cell i, and assume that the statistical office wants protection levels for the sensitive cell given by $LPL_{23} = 5$, $UPL_{23} = 8$ and $SPL_{23} = 0$.

We next give the main computations at each iteration, and later we will illustrate the construction of a violated in a particular separation call.

ITERATION 0: On this small example it is easy to implement an initial heuristic by following greedy ideas to find the solution represented in Figure 1 (b), so the initial upper bound on the optimal loss of information is 59. Nevertheless, to illustrate the algorithm described in the previous section the heuristic is not required.

Since the sensitive cell is located in row 2 and column 3, a constraint pool could be initialized with the following two constraints:

$$x_{43} + x_{13} + x_{33} \geq 1,$$

$$x_{24} + x_{21} + x_{22} \geq 1.$$

The initial master problem consists on the 15 variables (one associated to each potential secondary suppression) and the two cuts from the constraint pool. The initial solution is then:

$$x_{13}^* = 1, x_{21}^* = 1, x_{23}^* = 1,$$

hence the lower bound is 40. If the initial heuristic provided an upper bound of 59, then a problem reduction removes 10 variables because their reduced costs exceed $59 - 40 = 19$.

ITERATION 1: The separation procedures on the above solution find the following violated cuts:

$$x_{31} + x_{22} + x_{24} + x_{11} + x_{41} \geq 1,$$

$$x_{31} + x_{11} + x_{41} \geq x_{21},$$

$$x_{33} + x_{31} + x_{22} + x_{24} + x_{11} + x_{43} + x_{41} \geq x_{13}.$$

Then the new master problem contains 5 variables and 5 constraints, and the new solution is:

$$x_{13}^* = 1, x_{22}^* = 1, x_{23}^* = 1,$$

hence the lower bound is 51.

ITERATION 2: The separation procedures on the above solution find the following violated cuts:

$$x_{32} + x_{21} + x_{24} + x_{12} + x_{42} \geq 1,$$

$$x_{32} + x_{12} + x_{42} \geq x_{22},$$

$$x_{33} + x_{32} + x_{21} + x_{24} + x_{12} + x_{43} + x_{42} \geq x_{13}.$$

Then the new master problem contains 5 variables and 8 constraints, and the solution is:

$$x_{13}^* = 1, x_{21}^* = 1, x_{31}^* = 1, x_{23}^* = 1,$$

hence the lower bound is 57.

ITERATION 3: The separation procedures on the above solution find the following violated cuts:

$$x_{33} + x_{32} + x_{34} + x_{22} + x_{24} + x_{11} + x_{41} \geq 1,$$

$$x_{33} + x_{32} + x_{34} \geq x_{31},$$

$$x_{33} + x_{32} + x_{34} + x_{11} + x_{41} \geq x_{21},$$

$$x_{32} + x_{34} + x_{22} + x_{24} + x_{11} + x_{43} + x_{41} \geq x_{13}.$$

Then the new master problem contains 5 variables and 12 constraints, and the solution is:

$$x_{21}^* = 1, x_{23}^* = 1, x_{31}^* = 1, x_{33}^* = 1,$$

hence the lower bound is 59.

If we the initial heuristic found an upper bound of 59, then the procedure stops with the proof that the current heuristic solution is optimal. Otherwise, the procedure also stops after observing that the solution x^* is integer and satisfies all the protection requirements. On this simple example no cover inequality was generated and no branching was needed.

On this instance, each iteration introduces a few constraints since it solves only two linear problems. But on other instances with more sensitive cells and/or more attackers, more constraints are inserted in the master problem to impose all the protections levels requirements.

To illustrate the separation procedures, let us consider (for example) the solution x^* of the initial master problem (just before Iteration 1, i.e., $x_{13}^* = x_{21}^* = x_{23}^* = 1$). To check if there is a violated constraints, we solve two linear problems

$$\min \{y_{23} : (1) \text{ holds } \} \qquad \text{and} \qquad \max \{y_{23} : (1) \text{ holds } \}$$

with all the 16 variables y_i, since we are considering one sensitive cell and one attacker. (In general, the numbers of linear programs to be solved at each iteration is two times the number of sensitive times the number of attackers.)

The optimal value maximizing y_{23} is 22, which does not satisfy the upper protection requirement. Hence, we consider an optimal dual solution as for example:

$$\alpha_{11} = +1 \quad \text{(dual variable associated to } y_{11} \le 20\text{)}$$
$$\alpha_{12} = +1 \quad \text{(dual variable associated to } y_{12} \le 50\text{)}$$
$$\alpha_{43} = +1 \quad \text{(dual variable associated to } y_{43} \le 44\text{)}$$
$$\beta_{14} = +1 \quad \text{(dual variable associated to } y_{14} \ge 80\text{)}$$
$$\beta_{33} = +1 \quad \text{(dual variable associated to } y_{33} \ge 12\text{)}$$
$$\gamma_1 = -1 \quad \text{(dual variable associated to } y_{11} + y_{12} + y_{13} - y_{14} = 0\text{)}$$
$$\gamma_7 = +1 \quad \text{(dual variable associated to } y_{23} + y_{13} + y_{33} - y_{43} = 0\text{)}$$

Then, a violated constraint is

$$20x_{11} + 50x_{12} + 44x_{43} + 80x_{14} + 12x_{33} \ge 5$$

The associated strengthened constrains is

$$x_{11} + x_{12} + x_{43} + x_{14} + x_{33} \ge 1.$$

Of course, there are different optimal dual solutions, so there are different violated capacity constraints that can be detected by the separation procedure. In fact, from another optimal dual solution it follows the cut mentioned in Iteration 1.

By checking the lower protection level, it is possible to find another violated inequality. This second inequality could be different or same as the previous violated inequality. Of course, at each iteration only different inequalities must be considered.

7   A Relaxation of the General Problem

The multi-attacker CSP models presented in Section 4 require the resolution of two linear programming problems for each sensitive cell p and for each attacker k. From Computational Complexity Theory, the effort of solving all this family of problems is polynomial on $|I|$ and on $|P|$, hence it is not considered as a hard problem. Nevertheless, on large tables some simplifications could lead to a relaxed model and hence to a faster algorithm. Hopefully the simplifications do not come by reducing the protection requirements, thus the out-coming patterns remain still feasible. The cost of this speed up is paid on the suppression patterns that tend to be overprotected compared to an optimal pattern from the multi-attacker model.

A simplification of the problem known in literature consists on applying a classical CSP methodology (i.e., with one attacker) on an extended table where some sensitive cells have been aggregated into new dummy sensitive cells. By imposing larger protection level requirements on the dummy cells it is possible to avoid infeasible patterns that the classical CSP methodology could provided if applied directly to the original table. See, e.g., Robertson [15] for more details on this approach, where the dummy cells are called "unions" and they are built in practice by aggregating cells along a single axis of the table.

A different simplification consists on replacing all the internal intruders by one stronger attacker with all the a-priori information of each internal attacker. In other words, to reduce the computation effort required by the multi-attacker model, one idea could be to reduce the number of attackers by considering a fictitious worse-case attacker collecting all the a-priori information. Then the new model has a similar size to the one of the classical CSP, while it takes into account all knowledge from internal respondents. Therefore, a classical CSP methodology could be applied on the original table where the external bounds have been considerably reduced. This proposal was originally introduced by Jewett [9] using the concept of "protection capacity" associated to a cell and representing, when suppressed, its maximal contribution to disclosure another (sensitive) cell.

Of course, in this second simplification, a minor modification of the classical CSP must be introduced: By collecting all the individual a-priori information, a sensitive cell could have external bounds defining an range interval strictly contained into the range interval defined by its protection levels. Using the introductory example in Figure 1, the sensitive cell in Activity II and Region C could have an upper protection level of 8, a lower protection level of 10, and the external bounds defining the joint interval $[20, 25]$ from the information of their contributors. The classical CSP model without any modification would conclude that it is not possible to find any pattern with such protection requirements against an external attacker with such external bounds. Therefore, it is necessary to remove this simple check, i.e., in a classical CSP model (as defined in Section 2) we must replace

$$a_p - (a_p - lb_p) \leq y_p \leq a_p + (ub_p - a_p)$$

with

$$a_p - (a_p - lpl_p) \leq y_p \leq a_p + (upl_p - a_p)$$

in the definition of the attacker problems when protecting a sensitive cell p.

We call this new model the *intermediate CSP*, since the linear programs that must be solved for protecting each sensitive cell have much smaller feasible regions than the ones in the multiple attacker CSP, but bigger than the ones in the classical CSP. Indeed, first notice that by considering the intermediate CSP, "congruency" means also the knowledge of all the internal attackers, that are not present in the classical CSP. Second, the intermediate CSP solve, for each sensitive cell, a maximization (respectively minimization) linear program that coincides with the union of all the maximization (respectively minimization) linear programs than must be solved with the multiple attacker CSP problem, plus an equality for each cell i imposing that y_i is the same for all attackers.

	A B C	Total
Activity I	10 10 10	10
Activity II	0 0 0	10
Activity III	1 1 1	100
Total	10 10 10	190

Fig. 5. Loss of information for table in Figure 1

8   Relation between the Three Methodologies

Considering the multi-attacker model as the ideal one, the classical CSP on the original table could produce suppression patterns that are not protected, while the intermediate CSP could produce suppression pattern with over-protection. A clear example of the difference can be easily illustrated with the nominal table from Figure 1, where now cells (II, A), (II, B) and (II, C) are sensitives (perhaps because there is one respondent under each cell). Suppose also that the loss of information for each cell is defined by Figure 5. Assume that the required protection levels are all 1s, and the external bounds are 0 and $+\infty$ for suppressed cells.

By solving the multi-attacker CSP, the set of cell

$$\{(II, A), (II, B), (II, C), (III, A), (III, B), (III, C)\}$$

is a suppression pattern with minimal loss of information ($=3$). Notice that from such pattern, no internal attacker (and so also the external ones) will compute the exact value under a primary suppression. But, for the intermediate CSP, the previous pattern is not feasible, since when protecting cell (II, A) the attacker knows values under cells (II, B) and (II, C).

Hence, in general, the intermediate CSP provides over-suppressed patterns.

From this analysis, it is clear that the intermediate CSP is "intermediate" only on what concerns with effort to implement the algorithm, but not on what concerns the optimal loss of information. In fact, with respect to the multi-attacker CSP (the ideal version), the classical CSP produce smaller loss of information (and possibly infeasible patterns for internal attackers) and the intermediate CSP bigger loss of information (and feasible even for the worse case where all internal attackers joint their private information to disclose each sensitive cell).

9   Conclusion

We have introduced a mathematical description of the Cell Suppression Problem when the statistical office wants a suppression pattern minimizing the loss of information subject to protection level requirements for different sensitive cells and different attackers. Three mathematical models have been presented. A first one is in Bilevel Mathematical Programming. A second one is a first Integer

Linear Program that could be used on small/medium tables, and easily allow to impose integrality on the nominal unknown values. A third model is also in Integer Linear Programming, assumes that the nominal values are continuous numbers, and contains an smaller number of variables (so it could be also applied to large tables). The disadvantage of the number of constrains is solved by a modern cutting-plane generation. The main idea of this algorithm consists on an iterative procedure. At each iteration some constraints are considered whenever they are not satisfied. Even if the procedure analyzes whether the protection level requirements are satisfied for each sensitive cell and for each attacker, the overall procedure guarantee optimality for all the sensitive cells and all the attackers at the same time. In other words, the mechanism works with one cell and one attacker after another, in a sequential way, but the final effect is on all the cells and all the attackers. Therefore, the proposed approach does not only look for a good solution but an optimal one. The overall branch-and-cut algorithm was illustrated on the introductory instance.

We must remark the generality of the proposed model. In fact, it does not require any special structure on the table, but it can be applied to any k-dimensional, hierarchical or linked table. Moreover, there are no strong restriction on the extra knowledge of the attackers, since the model admits attackers with extra information related with more than one cells (which is the case in practice when a respondent contributes to several cell values).

For simplification in the notation, the description of the models and algorithms was done assuming that the extra a-priori information of each attacker is only on the bounds of the unknown values, i.e., the differences between attackers are only on the bounds in (1). Nevertheless, it is trivial to extend our model to work with more complex differences (if necessary in practice) between the information of the attackers more general than only on the external bounds. In fact, very similar models and algorithms can be designed when for each attacker there is an ad-hoc linear system of inequalities such as:

$$\left. \begin{array}{l} \sum_{i \in I} m_{ij}^k y_i \le b_j^k \qquad\qquad , j \in J^k \\ a_i - (a_i - lb_i^k)x_i \le y_i \le a_i + (ub_i^k - a_i)x_i \, , \, i \in I. \end{array} \right\}$$

where J^k is the set of inequalities known from attacker k. Of course, some of these inequalities are the ones defining the set of linear equations $[\sum_{i \in I} m_{ij} y_i \le b_j \, , \, j \in J]$ describing the table, but it is allowed to consider also additional ones known from attacker k. Therefore, the here-proposed model and algorithm can be easily adapted to protect a table against a set of a very wide variety of attackers. Indeed, our proposal manages any a-priori information as far as it can be formulated by a system of linear equalities/inequalities with integer/continuos variables. Obviously, there exists a-priori information that cannot be model as a linear system (for example, probabilistic knowledge), thus our proposal would not be able of finding optimal solution against attackers with such very special kind of information.

Finally, the here-presented proposal also allows the statistical office to impose different protection levels on different sensitive cells for each attacker to have a more accurate control on the desiderated protection. Therefore, the proposed model includes widely real requirements of the statistical office. The cost of this extra control is a model with a huge number of constraints, but the today state of the Mathematical Programming allows us to manage them efficiently on-the-fly.

This work finishes with an intermediate model between the well-known classical cell suppression and the new multi-attacker cell suppression. The intermediate model is an alternative to consider extra known information of several internal intruders by considering a single-attacker with all this information. Clearly this intermediate model requires the resolution of an smaller number of linear problems (two for each sensitive cell at each iteration), so an optimal solution could be found sooner. Nevertheless, this optimal solution would be protected for the worse-case attacker, hence it will be an overprotected pattern compared with an optimal pattern from the multi-attacker problem.

We are convinced that the availability of a common test-bed for comparing different solution procedures can be of great help to researchers in the field. Possibly in cooperation with EUROSTAT, we are planning to set-up and maintain on the Web CSPLIB, a test-problem library containing real-world (no longer confidential) and synthetic data available in the public domain. The library will include all tables along with the associated solutions, and will allow computational analysis of different techniques. Possible contributors can contact the author at *jjsalaza@ull.es* for more information; see the Appendix for the data format we are planning to use.

Acknowledgment

This work was inspired by a question of Anco Hundepool (*Central Bureau of Statistics*, The Netherlands) to the author during the "Second International Conference on Establishment Surveys", Buffalo, May 2000. We thank him and Sarah Giessing (*Federal Statistical Office*, Germany) for stimulating discussions on the topic. We also thank Giovanni Andreatta (*University of Padova*, Italy) for useful comments on this paper. Work supported by the European project IST-2000-25069, "Computational Aspects of Statistical Confidentiality" (CASC).

References

1. Cox, L.H. (1980) Suppression Methodology and Statistical Disclosure Control. *Journal of the American Statistical Association*, **75**, 377–385.
2. Cox, L.J. (1995) Network Models for Complementary Cell Suppression. *Journal of the American Statistical Association*, **90**, 1453–1462.
3. Carvalho, F.D., Dellaert, N.P., and Osório, M.S. (1994) Statistical Disclosure in Two-Dimensional Tables: General Tables. *Journal of the American Statistical Association*, **89**, 1547–1557.
4. Dellaert, N.P. and Luijten, W.A. (1996) Statistical Disclosure in General Three-Dimensional Tables. Technical Paper TI 96-114/9, Tinbergen Institute.
5. Fischetti, M. and Salazar, J.J. (1999) Models and Algorithms for the 2-Dimensional Cell Suppression Problem in Statistical Disclosure Control. *Mathematical Programming*, **84**, 283–312.

6. Fischetti, M. and Salazar, J.J. (2000) Models and Algorithms for Optimizing Cell Suppression Problem in Tabular Data with Linear Constraints. *Journal of the American Statistical Association*, **95**, 916–928.

7. Geurts, J. (1992) Heuristics for Cell Suppression in Tables. Technical Paper, Netherlands Central Bureau of Statistics, Voorburg.

8. Gusfield, D. (1988) A Graph Theoretic Approach to Statistical Data Security. *SIAM Journal on Computing*, **17**, 552–571.

9. Jewett, R. (1993) Disclosure Analysis for the 1992 Economic Census. Technical paper, U.S. Bureau of the Census, Washington.

10. Kao, M.Y. (1996) Data Security Equals Graph Connectivity. *SIAM Journal on Discrete Mathematics*, **9**, 87–100.

11. Kelly, J.P. (1990) Confidentiality Protection in Two and Three-Dimensional Tables. Ph.D. dissertation, University of Maryland, College Park, Maryland.

12. Kelly, J.P., Golden, B.L., and Assad, A.A. (1992) Cell Suppression: Disclosure Protection for Sensitive Tabular Data. *Networks* **22**, 397–417.

13. Nemhauser, G.L. and Wolsey, L.A. (1988) *Integer and Combinatorial Optimization*. New York: John Wiley & Sons.

14. Robertson, D.A. (1994) Cell Suppression at Statistics Canada. Proceedings of the *Second International Conference on Statistical Confidentiality*, Luxembourg.

15. Robertson, D.A. (2000) Improving Statistics Canada's cell suppression software (CONFID). Technical paper, Statistics Canada, Ottawa, Canada.

16. Sande, G. (1984) Automated cell suppression to preserve confidentiality of business statistics. *Statistical Journal of the United Nations ECE* **2**, 33–41.

17. Sande, G. (1995) ACS documentation. *Sande & Associates*, 600 Sanderling Ct. Secaucus NJ, 07094 U.S.A.

18. Willenborg, L.C.R.J. and De Waal, T. (1996) *Statistical Disclosure Control in Practice. Lecture Notes in Statistics 111*. New York: Springer.

Appendix: Input Data Format for CSPLIB Instances

CSPLIB instances are described by ASCII files. Two input formats are possible, one for k-dimensional tables with marginals and $1 \leq k \leq 4$, and another for more general tables with entries linked by linear equations. The second format allows one to also describe k-dimensional tables with marginals, for any k. For both formats, an example is given which refers to the following 2-dimensional table with marginals:

1	**3**	**5**	7	16
10	12	**11**	22	42
8	**6**	4	**2**	20
19	**21**	20	**18**	78

The input format for a k-dimensional table $(1 \leq k \leq 4)$ starts with the INTEGER k and another INTEGER number l for the number of attackers. Then k INTEGER numbers give the table size for each dimension.

A list of lines follows, each associated with a different table cell and containing the following $k + 3 + 5 \times l$ items:

- the first k items are INTEGER numbers identifying the index position of the cell (0 for marginal indices);
- a REAL number representing the nominal value of the cell;
- an INTEGER number representing the loss of information incurred if the cell is suppressed;
- a CHARACTER representing the status of the cell ('u' for a sensitive cell, 'z' for enforcing cell publication; 's' for a potential complementary suppression);
- a REAL number representing the lower bound of the cell known to the attacker;
- a REAL number representing the upper bound of the cell known to the attacker;
- a REAL number representing the lower protection level of the cell required by the statistical office;
- a REAL number representing the upper protection level of the cell required by the statistical office;
- a REAL number representing the sliding protection level of the cell required by the statistical office.

The last 5 fields must be repeated in each line, one describing the features of each attacker.

Here is the input file for the sample instance, assuming two intruders. One has the external bounds of 0 and 100 on all unpublished cells, while the other has the extra knowledge that cell (1,3) is between 3 and 10, so the statistical office requires a bigger upper protection level (i.e., 3 instead of 1).

```
========================================= top-of-file
2 2
3 4
0 0 78.0  4 s   0.0 100.0  0.0  0.0  0.0    0.0 100.0  0.0  0.0  0.0
0 1 19.0  2 s   0.0 100.0  0.0  0.0  0.0    0.0 100.0  0.0  0.0  0.0
0 2 21.0  0 u   0.0 100.0  2.0  4.0  0.0    0.0 100.0  2.0  4.0  0.0
0 3 20.0  2 s   0.0 100.0  0.0  0.0  0.0    0.0 100.0  0.0  0.0  0.0
0 4 18.0  0 u   0.0 100.0  1.0  3.0  0.0    0.0 100.0  1.0  3.0  0.0
1 0 16.0  2 s   0.0 100.0  0.0  0.0  0.0    0.0 100.0  0.0  0.0  0.0
1 1  1.0  0 u   0.0 100.0  0.0  4.0  0.0    0.0 100.0  0.0  4.0  0.0
1 2  3.0  0 u   0.0 100.0  1.0  3.0  0.0    0.0 100.0  1.0  3.0  0.0
1 3  5.0  0 u   0.0 100.0  0.0  1.0  0.0    3.0  10.0  0.0  3.0  0.0
1 4  7.0  1 s   0.0 100.0  0.0  0.0  0.0    0.0 100.0  0.0  0.0  0.0
2 0 42.0  2 s   0.0 100.0  0.0  0.0  0.0    0.0 100.0  0.0  0.0  0.0
2 1 10.0  0 u   0.0 100.0  1.0  4.0  0.0    0.0 100.0  1.0  4.0  0.0
2 2 12.0  1 s   0.0 100.0  0.0  0.0  0.0    0.0 100.0  0.0  0.0  0.0
2 3 11.0  0 u   0.0 100.0  2.0  3.0  0.0    0.0 100.0  2.0  3.0  0.0
2 4  9.0  1 s   0.0 100.0  0.0  0.0  0.0    0.0 100.0  0.0  0.0  0.0
3 0 20.0  2 s   0.0 100.0  0.0  0.0  0.0    0.0 100.0  0.0  0.0  0.0
3 1  8.0  1 s   0.0 100.0  0.0  0.0  0.0    0.0 100.0  0.0  0.0  0.0
3 2  6.0  0 u   0.0 100.0  2.0  0.0  0.0    0.0 100.0  2.0  0.0  0.0
3 3  4.0  1 s   0.0 100.0  0.0  0.0  0.0    0.0 100.0  0.0  0.0  0.0
3 4  2.0  0 u   0.0 100.0  0.0  4.0  0.0    0.0 100.0  0.0  4.0  0.0
========================================= end-of-file
```

The input format for a linearly constrained table is organized into two sections: the first section contains the cell description, while the second contains the equation description. Moreover, it is required that the first item of the file is '0' (i.e., INTEGER zero) and then it follows an INTEGER number with the number l of attackers.

The cell description section starts with the INTEGER number n of cells. It follows n lines containing $4 + 5 \times l$ items each:

- an INTEGER number from 0 to $n - 1$ with the cell index;
- a REAL number representing the nominal value of the cell;
- an INTEGER number representing the loss of information incurred if the cell is suppressed;
- a CHARACTER representing the status of the cell ('u' for a sensitive cell, 'z' for enforcing cell publication; 's' for a potential complementary suppression);
- a REAL number representing the lower bound of the cell known to the attacker;
- a REAL number representing the upper bound of the cell known to the attacker;
- a REAL number representing the lower protection level of the cell required by the statistical office;
- a REAL number representing the upper protection level of the cell required by the statistical office;
- a REAL number representing the sliding protection level of the cell required by the statistical office.

The last 5 fields must be repeated in each line, one describing the features of each attacker.

The equation description section starts with the INTEGER number m of equations, followed by m lines. Each line starts with two numbers, a first REAL number giving the right-hand-side value of the correspondent equation, and a second INTEGER number k giving the number of variables in its left-hand side. The line continues with the separator ':', followed by k pairs of the form i (c_i), where the INTEGER number i gives the cell index and the INTEGER number c_i gives the corresponding nonzero left-hand-side coefficient in the equation.

Here is the input file for the sample instance.

```
=============================================== top-of-file
0 2
20
 0 78.0  4 s    0.0 100.0  0.0  0.0  0.0    0.0 100.0  0.0  0.0  0.0
 1 19.0  2 s    0.0 100.0  0.0  0.0  0.0    0.0 100.0  0.0  0.0  0.0
 2 21.0  0 u    0.0 100.0  2.0  4.0  0.0    0.0 100.0  2.0  4.0  0.0
 3 20.0  2 s    0.0 100.0  0.0  0.0  0.0    0.0 100.0  0.0  0.0  0.0
 4 18.0  0 u    0.0 100.0  1.0  3.0  0.0    0.0 100.0  1.0  3.0  0.0
 5 16.0  2 s    0.0 100.0  0.0  0.0  0.0    0.0 100.0  0.0  0.0  0.0
 6  1.0  0 u    0.0 100.0  0.0  4.0  0.0    0.0 100.0  0.0  4.0  0.0
 7  3.0  0 u    0.0 100.0  1.0  3.0  0.0    0.0 100.0  1.0  3.0  0.0
 8  5.0  0 u    0.0 100.0  0.0  1.0  0.0    3.0  10.0  0.0  3.0  0.0
```

```
 9  7.0  1 s   0.0 100.0  0.0  0.0  0.0    0.0 100.0  0.0  0.0  0.0
10 42.0  2 s   0.0 100.0  0.0  0.0  0.0    0.0 100.0  0.0  0.0  0.0
11 10.0  0 u   0.0 100.0  1.0  4.0  0.0    0.0 100.0  1.0  4.0  0.0
12 12.0  1 s   0.0 100.0  0.0  0.0  0.0    0.0 100.0  0.0  0.0  0.0
13 11.0  0 u   0.0 100.0  2.0  3.0  0.0    0.0 100.0  2.0  3.0  0.0
14  9.0  1 s   0.0 100.0  0.0  0.0  0.0    0.0 100.0  0.0  0.0  0.0
15 20.0  2 s   0.0 100.0  0.0  0.0  0.0    0.0 100.0  0.0  0.0  0.0
16  8.0  1 s   0.0 100.0  0.0  0.0  0.0    0.0 100.0  0.0  0.0  0.0
17  6.0  0 u   0.0 100.0  2.0  0.0  0.0    0.0 100.0  2.0  0.0  0.0
18  4.0  1 s   0.0 100.0  0.0  0.0  0.0    0.0 100.0  0.0  0.0  0.0
19  2.0  0 u   0.0 100.0  0.0  4.0  0.0    0.0 100.0  0.0  4.0  0.0
9
0.0  5 : 0 (-1) 1 (1) 2 (1) 3 (1) 4 (1)
0.0  5 : 5 (-1) 6 (1) 7 (1) 8 (1) 9 (1)
0.0  5 : 10 (-1) 11 (1) 12 (1) 13 (1) 14 (1)
0.0  5 : 15 (-1) 16 (1) 17 (1) 18 (1) 19 (1)
0.0  4 : 0 (-1) 5 (1) 10 (1) 15 (1)
0.0  4 : 1 (-1) 6 (1) 11 (1) 16 (1)
0.0  4 : 2 (-1) 7 (1) 12 (1) 17 (1)
0.0  4 : 3 (-1) 8 (1) 13 (1) 18 (1)
0.0  4 : 4 (-1) 9 (1) 14 (1) 19 (1)
========================================= end-of-file
```

Network Flows Heuristics for Complementary Cell Suppression: An Empirical Evaluation and Extensions*

Jordi Castro**

Statistics and Operations Research Dept., Universitat Politècnica de Catalunya
Pau Gargallo 5, 08028 Barcelona (Spain)
jcastro@eio.upc.es
http://www-eio.upc.es/ jcastro

Abstract. Several network flows heuristics have been suggested in the past for the solution of the complementary suppression problem. However, a limited computational experience using them is reported in the literature, and, moreover, they were only appropriate for two-dimensional tables. The purpose of this paper is twofold. First, we perform an empirical comparison of two network flows heuristics. They are improved versions of already existing approaches. Second, we show that extensions of network flows methods (i.e., multicommodity network flows and network flows with side constraints) can model three-dimensional, hierarchical and linked tables. Exploiting this network structure can improve the performance of any solution method solely based on linear programming formulations.

Keywords: Complementary cell suppression problem, linear programming, network optimization, network flows with side constraints, multicommodity network flows.

1 Introduction

Cell suppression is a widely used technique by statistical agencies to avoid the disclosure of confidential tabular data. Given a list of primary cells to be protected, the objective of the cell suppression problem (CSP) is to find a set of complementary cells that have to be additionally suppressed. This pattern of suppressions is found under some criteria as, e.g., minimum number of suppressions, or minimum value suppressed.

CSP was shown to be NP-hard in [19]. This motivated that most of the former approaches focused on heuristic methods for approximate solutions. Methods based on graph theory were suggested for two-dimensional tables in [4], and extended for three dimensions in [8]; they were designed for general tables. The hypercube method, currently under consideration by the Federal Statistical Office of Germany, was also based in geometric considerations of the problem [14].

---

* Work supported by the IST-2000-25069 CASC project.
** Author supported by CICYT Project TAP99-1075-C02-02.

J. Domingo-Ferrer (Ed.): Inference Control in Statistical Databases, LNCS 2316, pp. 59–73, 2002.
© Springer-Verlag Berlin Heidelberg 2002

A different kind of techniques was obtained by using linear programming (LP), in particular, network optimization. This paper is devoted to this kind of methods. We will perform an empirical evaluation of two network flows heuristics. It will also be shown how three-dimensional, linked and hierarchical tables can be modeled as multicommodity network flows and network flows with side constraints [1]. This allows the use of specialized LP codes that exploit the network structure of the problem [5,6,13].

There is a fairly extensive literature on network flows methods for CSP. In [19] an algorithm for sliding protection was suggested, but it was only applied to small scale two-dimensional tables. In [7] several alternative network methods were reviewed, some of them successfully applied in practice in U.S. [18] and Canada [21] (although in this latter case a pure LP formulation was considered). However, these heuristics were only appropriate for two-dimensional tables, since they relied on a minimum cost network flows solver. Multi-dimensional, linked and hierarchical tables had to be split into several two-dimensional tables, which forced some kind of backtracking procedure. This inconvenient could be removed by using general LP solvers. However, limitations of past LP technology resulted in inefficient implementations [18]. As noted in [3] this drawback has been overcome by current LP solvers, and some steps have been performed for including them in CSP production codes [20]. Exploiting that three-dimensional, linked and hierarchical tables can be modeled through multicommodity networks and networks with side constraints opens the possibility of using a new range of solvers for CSP.

Recently, an exact procedure based on state-of-the-art mixed integer linear programming (MILP) techniques (i.e., branch-and-cut and Bender's decomposition) was proposed in [9,10]. This method has been able to solve large non-trivial CSP instances very efficiently. As stated in [10], the exact algorithm developed includes an initial LP-based heuristic phase, which is similar to those suggested in the past using network flows codes. Therefore, the exact procedure can also take profit of recent improvements in heuristic methods. Moreover, the Bender's decomposition subproblems could also benefit of the underlying network with side constraints structure of the constraints matrix.

This paper is organized as follows. Section 2 shows the exact formulation and a linear relaxation of CSP. The two heuristics considered in this work are based on the exact and relaxed formulations, respectively. These heuristics are outlined in Section 3 and compared in Section 4. Section 5 shows the extension of network flows models for three-dimensional, linked and hierarchical tables. Finally, Section 6 presents some preliminary computational results with a network flows heuristic for three-dimensional tables.

2 Formulation of CSP

Given a (usually) positive table (i.e., a set of cells $a_i \geq 0, i = 1 \ldots \bar{n}$, satisfying some linear relations $Aa = b$), a set $\mathcal{P}$ of $|\mathcal{P}|$ primary cells to be protected, and upper and lower protection levels U_i and L_i for each primary cell $i = 1 \ldots |\mathcal{P}|$, the purpose of CSP is to find a set $\mathcal{C}$ of additional complementary cells whose suppression guarantees that, for each $p \in \mathcal{P}$,

$$\underline{a_p} \leq a_p - L_p \quad \text{and} \quad \overline{a_p} \geq a_p + U_p, \tag{1}$$

$\underline{a_p}$ and $\overline{a_p}$ being defined as

$$
\begin{array}{llll}
\underline{a_p} = \displaystyle\min_{x_i, i=1...\overline{n}} x_p & & \overline{a_p} = \displaystyle\max_{x_i, i=1...\overline{n}} x_p & \\
\text{s.t.} & Ax = b & \text{s.t.} & Ax = b \\
& x_i \geq 0 \ \ i \in \mathcal{P} \cup \mathcal{C} & \text{and} & x_i \geq 0 \ \ i \in \mathcal{P} \cup \mathcal{C} \\
& x_i = a_i \ \ i \notin \mathcal{P} \cup \mathcal{C} & & x_i = a_i \ \ i \notin \mathcal{P} \cup \mathcal{C}.
\end{array} \tag{2}
$$

$\underline{a_p}$ and $\overline{a_p}$ in (2) are the lowest and greatest possible values that can be deduced for each primary cell from the published table, once the entries in $\mathcal{P} \cup \mathcal{C}$ have been suppressed. Imposing (1), the desired level of protection is guaranteed. CSP can thus be formulated as an optimization problem of minimizing some function that measures the cost of suppressing additional cells subject to that conditions (1) and (2) are satisfied for each primary cell.

CSP was first formulated in [19] as a large MILP problem. For each entry a_i a binary variable $y_i, i = 1 \ldots \overline{n}$ is considered. y_i is set to 1 if the cell is suppressed, otherwise is 0. For each primary cell $p \in \mathcal{P}$, two auxiliary vectors $x^{l,p} \in \mathbb{R}^{\overline{n}}$ and $x^{u,p} \in \mathbb{R}^{\overline{n}}$ are introduced to impose, respectively, the lower and upper protection requirements of (1) and (2). These vectors represent cell deviations (positive or negative) from the original a_i values. The resulting model is

$$
\begin{aligned}
\min \quad & \sum_{i=1}^{\overline{n}} a_i y_i \\
\text{s.t.} \quad & \left.
\begin{array}{l}
Ax^{l,p} = 0 \\
-a_i y_i \leq x_i^{l,p} \leq M y_i \quad i = 1 \ldots \overline{n} \\
x_p^{l,p} = -L_p \\[2mm]
Ax^{u,p} = 0 \\
-a_i y_i \leq x_i^{u,p} \leq M y_i \quad i = 1 \ldots \overline{n} \\
x_p^{u,p} = U_p
\end{array}
\right\} \quad \text{for each } p \in \mathcal{P} \\[2mm]
& y_i \in \{0,1\}
\end{aligned} \tag{3}
$$

Inequality constraints impose the bounds of $x_i^{l,p}$ and $x_i^{u,p}$ when $y_i = 1$ (M being a large value), and prevent deviations in nonsuppressed cells (i.e., $y_i = 0$). Clearly, the constraints of (3) guarantee that the solutions of the linear programs (2) will satisfy (1). A similar formulation was used in [10].

The first heuristic of Section 3 was inspired by the above formulation, since it attempts to find a "good" (i.e., close to the optimum) feasible point for (3) considering the combinatorial nature of the objective function. However, most network flows heuristics consider a linear objective function. They can thus be seen as derived from a linear relaxation of (3). This linear relaxation can be obtained by replacing the binary variables y_i in (3) by two variables z_i^l and z_i^u, $i = 1 \ldots \overline{n}$, that represent the minimum deviation required in a cell to guarantee, respectively, the lower and upper protection levels of the primary cells. This linear relaxation model can be stated as

$$
\begin{aligned}
& \min \quad \sum_{i=1}^{\overline{n}} a_i(z_i^l + z_i^u) \\
& \text{s.t.}
\end{aligned}
$$

$$
\left. \begin{aligned}
Ax^{l,p} &= 0 \\
-z_i^l \leq x_i^{l,p} &\leq z_i^u \qquad i = 1 \ldots \overline{n} \\
x_p^{l,p} &= -L_p \\[4pt]
Ax^{u,p} &= 0 \\
-z_i^l \leq x_i^{u,p} &\leq z_i^u \qquad i = 1 \ldots \overline{n} \\
x_p^{u,p} &= U_p
\end{aligned} \right\} \quad \text{for each } p \in \mathcal{P} \qquad (4)
$$

$$
z^l, z^u \geq 0
$$

The term "partial cell suppression problem" was coined in [11] for (4).

As noted in [10], (3) and (4) give rise to very large MILP and LP problems even for tables of moderate sizes and number of primary cells. However, their constraints matrices are highly structured. For instance, Figure 1 shows the structure of the constraints matrix for problem (4). This dual-block structure is similar to that of stochastic programming problems [2]. Such structure can be exploited through decomposition schemes, as done in [10] using Bender's method. As noted in [2], alternative decomposition approaches based on interior-point methods [16] could be attempted, although, in principle, and from the computational experience reported in [10], they don't look like a promising choice for CSP.

The second level of structure in (3) and (4) comes from the table linear relations matrix A. This structure has only been exploited up to now for two-dimensional tables, which can be modeled as a bipartite network. The two heuristics described in the next section belong to this category. In Section 5 it will be shown how more complicated structures can also be exploited through extended network flows formulations.

3   Network Flows Heuristics

Network flows heuristics attempt to find approximate solutions to (3) or (4). In fact, they can only guarantee a (hopefully good) feasible solution, i.e., a pattern of complementary suppressions $\mathcal{C}$ satisfying the lower and upper protection levels. Before outlining the two approaches considered in this work, we present the framework for CSP network flows heuristics. In this section we will mainly focus on two-dimensional tables.

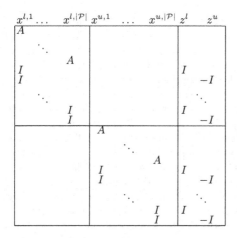

Fig. 1. Constraints matrix structure for problem (4)

3.1 General Framework

Heuristics for two-dimensional CSPs exploit that the linear relations of a $(m + 1) \times (n + 1)$ table defined by system $Ax = b$ can be modeled as the network of Figure 2. Arcs are associated to cells and nodes to equations; row $m + 1$ and column $n + 1$ correspond to marginals. For each cell two variables x_i^+ and x_i^- are defined, denoting respectively a positive or negative deviation from the cell value a_i. Clearly, the feasible deviations must satisfy

$$A(x^+ + x^-) = 0, \tag{5}$$

which can be modeled as a nonoriented version of the network of Figure 2.

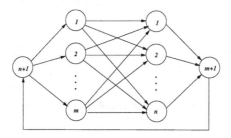

Fig. 2. Network representation of a $(m + 1) \times (n + 1)$ table

CSP heuristics perform one iteration for each primary cell $p \in \mathcal{P}$, as shown in Figure 3. At each iteration the protection levels of p are satisfied through the solution of one or two successions of network flows problems. Some heuristics (e.g., that of Section 3.3) only need to solve a single network flows problem for each succession, while others (e.g., that of Section 3.2) can require a few ones to guarantee the protection levels. This is stated at line 5 of Figure 3. Two (successions of) network flows problems are solved when we allow different patterns of complementary suppressions for the upper and lower protection levels of p. This is in principle the best option, since it is closer to the original formulations of CSP (3) and (4) (where different vectors $x^{l,p}$ and $x^{l,p}$ were also considered). On the other hand, one (succession of) network flows problem(s) is solved when we look for a pattern of suppressions that satisfies both the upper and lower protection levels. This was the choice in [18]. Although the maximum number of network flows problems to be solved halves, this heuristic can not be considered a good approximation to (3) and (4), and therefore, in theory, it should not be able to provide good solutions. As in [19] and the heuristic method of [10], we considered the two-networks-flows (successions) approach, as shown at line 3 of Figure 3. The structure of the network flows problems solved at line 6 of the algorithm (i.e., injections, costs, and lower and upper capacities) depends on the particular heuristic being used.

Algorithm *CSP Heuristic Framework(Table,$\mathcal{P}, U, L$)*
1    $\mathcal{C} = \emptyset$; $CLP_i = 0$, $CUP_i = 0$, $i \in \mathcal{P}$;
2   **for_each** $p \in \mathcal{P}$ **do**
3   **for_each** type of protection level $X \in \{U, L\}$ **do**
4   **if** $CXP_p < X_p$ ($X = L$ or $X = U$) **then**
5   **repeat** /* some heuristics only require one repeat iteration */
6   Solve network problem with flows x^+ and x^-;
7   Obtain set $\mathcal{T} = \{i : x_i^+ + x_i^- > 0\}$ of positive flows;
8               $\mathcal{C} := \mathcal{C} \cup \mathcal{T} \setminus \mathcal{P}$;
9   Update current protection levels CXP ($X = L$ or $X = U$);
10   **until** protection level X_p is achieved ($X = L$ or $X = U$);
11   **end_if**
12   **end_for_each**
13 **end_for_each**
End_algorithm

Fig. 3. General framework of heuristic procedures

After the solution of the network flows problem, additional suppressions are obtained, which are added to the current set $\mathcal{C}$ of complementary cells (lines 7 and 8 of Figure 3). The new suppressions can also satisfy the protection levels of the following primary cells. To avoid the solution of unnecessary network flows problems, we maintain two vectors CLP_i and CUP_i, $i \in \mathcal{P}$, with respectively the current lower and upper protection achieved for all primary cell. Primary cell p is thus not treated if its protection levels are satisfied (line 4). This is a significant improvement of the methods described in [7,19]. The particular updating of CLP and CUP at line 9 of the algorithm is heuristic dependent. It is noteworthy that the algorithm of Figure 3 is also valid for three-dimensional, linked, and hierarchical tables. We only need to substitute the optimization problem at line 6 of the algorithm.

3.2   First Heuristic

The first heuristic is derived from that presented in [7]. This heuristic uses a variant of the objective function of (3), i.e., a suppressed cell has a fixed cost. To this end, only 0-1 values are allowed for the variables (arcs) x^+ and x^- that flow through the nonoriented network of Figure 2. Due to the unimodularity of the bases of network simplex methods [1], this is guaranteed if we impose bounds $0 \leq x^+ \leq 1$ and $0 \leq x^- \leq 1$. The objective function of the network problem $\sum_{i=1}^{\overline{n}} c_i(x_i^+ + x_i^-)$ (c_i being discussed later) is then a good approximation of that of (3).

We force a flow of 1 through arc x_p^+, p being the current primary cell selected at line 2 of Figure 3. This can be done either imposing a lower bound of 1 or using a high negative cost for x_p^+. The upper bound of x_p^- is set to 0 to avoid a trivial cycle. After solving the network flows problem, a cycle $\mathcal{T}$ of cells with 1-flows will be obtained and added to the set of complementary suppressions according to lines 7–8 of Figure 3. If the value $\gamma = \min\{a_i : i \in \mathcal{T}\}$ is greater than X_p ($X = L$ or $X = U$, following the notation of Figure 3), then the (lower or upper) protection level of p is guaranteed. If $\gamma < X_p$, we have only partially protected cell p; we then set $X_p := X_p - \gamma$ and repeat the procedure, finding additional cycles until the protection level is satisfied. This iterative procedure corresponds to lines 5–10 of Figure 3. In this heuristic only a vector of current protection level is maintained (i.e., $CLP_i = CUP_i$ in Figure 3), since the γ value can be used for both the upper and lower protection of the cells in the cycle.

The behavior of the heuristic is governed by the costs c_i of variables x_i^+ and x_i^- associated to cells a_i. Costs are chosen to force the selection of, first, cells $\in \mathcal{P} \cup \mathcal{C}$ and $a_i \geq X_p$, second, cells $\notin \mathcal{P} \cup \mathcal{C}$ and $a_i \geq X_p$, third, cells $\in \mathcal{P} \cup \mathcal{C}$ and $a_i < X_p$, and, finally, cells $\notin \mathcal{P} \cup \mathcal{C}$ and $a_i < X_p$ ($X = L$ or $X = U$), in an attempt to balance the number of new complementary suppressions and network flows problems to be solved. Clearly, for each of the above four categories, cells with the lowest a_i values are preferred.

The above network problem can also be formulated as a shortest-path between the row and column nodes of primary cell p in the nonoriented network of Figure 2, which relates this heuristic with the method described in [4] for general tables.

3.3   Second Heuristic

The second approach is based on [19] and is similar to the heuristics used by the U.S. Census Bureau [18] and Statistics Canada [21], and the heuristic of

[10]. Unlike the original method, that solved a single network flows problem [19], two separate problems are considered for the lower and upper protection of each primary cell (line 4 of Figure 3). For both problems, we set bounds $x_i^+ \geq 0$ and $a_i \geq x_i^- \geq 0$. For lower protection we force a flow L_p through arc x_p^-, while a flow U_p is sent through arc x_p^+ in the upper protection problem. Unlike the heuristic of Section 3.2, only one network flows problem needs to be solved for each protection level (lines 5 and 10 of Figure 3 are no longer required). The objective function $\sum_{i=1}^{\overline{n}} c_i(x_i^+ + x_i^-)$, $c_i = 0$ if $i \in \mathcal{P} \cup \mathcal{C}$ and $c_i = a_i$ otherwise, is minimized subject to (5) for each protection level. This objective is related with that of (4).

As in the first heuristic, after the solution of each network flows problem we obtain the cycle $\mathcal{T}$ and update the complementary suppressions (lines 7 and 8 of Figure 3). Defining $\gamma = \min\{a_i : i \in \mathcal{T}\}$, the current protection vectors are updated at line 9 of Figure 3 as $CLP_i := \max\{CLP_i, \gamma, x_i^- - x_i^+\}$ after the solution of the lower protection problem, and $CUP_i := \max\{CUP_i, \gamma, x_i^+ - x_i^-\}$ after solving the upper protection problem, for all $i \in \mathcal{T}$.

The lower-bounding procedure described in [19] was also applied to obtain an initial set of complementary suppressions. The clean-up post-process suggested in [19] to remove unnecessary complementary suppressions was not performed since it is computationally very inefficient.

4   Computational Comparison

The heuristics of Sections 3.2 and 3.3 have been implemented with the AMPL modeling language [12], which allows the quick development of algorithm prototypes. The network flows problems were solved with the Cplex 6.5 network simplex code [17]. Two generators for two-dimensional positive tables were developed. The first generator follows the description of [19]. Cell values are randomly obtained from an integer uniform distribution [1,1000] with probability 0.8 and are 0 with probability 0.2. The second one is similar to generator 1 of [9]. Cell values are randomly obtained from integer uniform distributions [1,4] for primary cells and $\{0\} \cup [5, 500]$ for the remaining entries. Primary cells are randomly chosen from the internal cells in both generators.

We produced 48 instances with each generator. Each instance is defined by three parameters (m, n, p), which denote the number of rows, columns and primary suppressions, respectively. The 48 instances were obtained considering all the combinations for $m \in \{50, 100, 150\}$ and $n, p \in \{50, 100, 150, 200\}$. In all the cases the lower and upper protection levels were a 15% of the cell value. This symmetric range protection slightly benefits the heuristic of Section 3.2, because one single network flows problem is enough for both the upper and lower protection requirements.

The results obtained are shown in Figures 4–11. The heuristics of Sections 3.2 and 3.3 are denoted as "first" and "second" heuristic, respectively. Executions were carried out on a Sun Ultra2 200MHz workstation (approximately equivalent in performance to a 350 Mhz Pentium PC). The horizontal axes of all the figures refer to the instance number, $(50, 50, 50)$ being the first, $(50, 50, 100)$ the second and so on. The several groups of four points with positive slope correspond to the same table size for different numbers of primary suppressions.

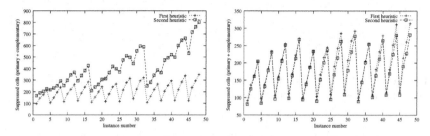

Fig. 4. Number of suppressed cells for instances of generator 1

Fig. 5. Number of suppressed cells for instances of generator 2

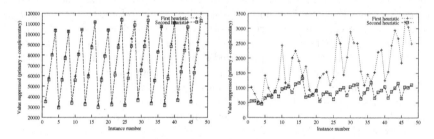

Fig. 6. Total value suppressed for instances of generator 1

Fig. 7. Total value suppressed for instances of generator 2

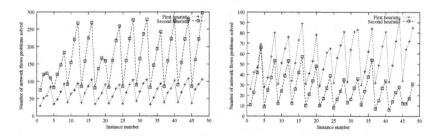

Fig. 8. Network flows problems solved for instances of generator 1

Fig. 9. Network flows problems solved for instances of generator 2

Clearly, the behavior of the heuristics depends on the set of instances. The first heuristic is more efficient for instances obtained with generator 1, since it suppresses less cells, solves less network flows problems, and is faster than the

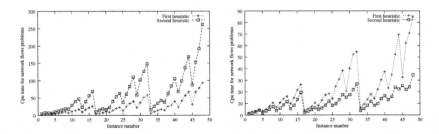

Fig. 10. CPU time of network flows problems for instances of generator 1

Fig. 11. CPU time of network flows problems for instances of generator 2

second heuristic (Figures 4, 8 and 10 respectively). However, the total value suppressed is similar, as shown in Figure 6. On the other hand, the second heuristic provides slightly better results for the second set of instances, mainly for the total value suppressed, number of network flows problems solved, and execution time (Figures 7, 9 and 11 respectively). Therefore, the choice of a heuristic should consider the particular structure of the tables to be protected.

It must be noted that, in theory, execution times could be improved for the first heuristic if we used some specialized shortest-path algorithm, instead of a minimum cost network solver. Unfortunately, this heuristic can not be easily extended to three-dimensional, linked and hierarchical tables. As shown in next Section, these tables can be modeled through multicommodity flows and flows with side constraints, which do not preserve the integrality property of minimum cost network flows models.

5   Extensions of Network Flows Models

In previous works on CSP, the structure of the table linear relations matrix A was only exploited for two-dimensional tables, which, as shown in Section 3, were modeled as a bipartite network. For more complicated structures, A was considered as the constraints matrix of a general LP problem [10,18]. However extensions of network models (i.e., multicommodity flows, and flows with side constraints) can still be applied for three-dimensional, linked and hierarchical tables.

Multicommodity flows are a generalization of network flows problems. In these models l commodities have to be routed through the same underlying network. The set of feasible multicommodity flows is

$$\mathcal{F}_1 = \{(x^1, \ldots, x^l) : Nx^k = b^k, l^k \leq x^k \leq u^k, k = 1 \ldots l, \sum_{k=1}^{l} x^k \leq u\}, \quad (6)$$

where x^k is the flows vector for each commodity $k = 1 \ldots l$, N is the node-arc incidence network matrix, b^k are the node injections at the network for each commodity, u^k and l^k are respectively the individual lower and upper arc capacities for each commodity, and u is the mutual arc capacity for all the commodities. This kind of models have been extensively applied in distribution, routing, logistic and telecommunications problems [1].

In networks with side constraints the flows of some arcs must satisfy additional linear relations. Mutual capacity constraints of multicommodity problems are a particular case of side constraints. The set of feasible flows is thus defined as

$$\mathcal{F}_2 = \{x : Nx = b, \underline{b} \leq Tx \leq \overline{b}, l \leq x \leq u\}, \tag{7}$$

x being the flows vector, N the node-arc incidence network matrix, b the injections at the nodes of the network, u and l the lower and upper capacities of the arcs, and T the side constraints matrix.

Although the minimization of a linear cost function subject to constraints (6) or (7) can be solved with a general algorithm for LP, several specialized methods that exploit the partial network structure have been developed. Among them we find simplex-based [6], Lagrangian relaxation [13], interior-point [5] and approximation methods [15]. Since multicommodity flows and flows with side constraints can be used to model three-dimensional, linked and hierarchical tables, as shown in next subsections, these algorithms can also be applied to CSP. Moreover, unlike most efficient LP solvers, some of these specialized algorithms are freely available for noncommercial purposes (e.g., [5,6]).

5.1   Three-Dimensional Tables

The linear relations $Aa = b$ of the cell values a_i of a $(m+1) \times (n+1) \times (l+1)$ three-dimensional table can be stated as

$$\sum_{i=1}^{m} a_{ijk} = a_{(m+1)jk} \quad j = 1 \ldots n, \quad k = 1 \ldots l \tag{8}$$

$$\sum_{j=1}^{n} a_{ijk} = a_{i(n+1)k} \quad i = 1 \ldots m, \quad k = 1 \ldots l \tag{9}$$

$$\sum_{k=1}^{l} a_{ijk} = a_{ij(l+1)} \quad i = 1 \ldots m, \quad j = 1 \ldots n. \tag{10}$$

Cells $a_{(m+1)jk}$, $a_{i(n+1)k}$ and $a_{ij(l+1)}$ form, respectively, the row-marginal $n \times l$ table, the column-marginal $m \times l$ table, and the level-marginal $m \times n$ table.

Clearly, putting together in (8) and (9) equations related to the same level k, (8)–(10) can be written as

$$Na^k = b^k, \quad k = 1 \ldots l \tag{11}$$

$$\sum_{k=1}^{l} a^k = a^{l+1}, \tag{12}$$

N being the network linear relations of the two-dimensional table associated to each level (depicted in Figure 2), a^k the $m \times n$ cells (flows) of level k, b^k the row and column marginals of level k, and a^{l+1} the level marginal values. From (6), it is clear that (11) and (12) define a set of feasible multicommodity flows, by choosing appropriate upper and lower bounds u^k and l^k (e.g., $u^k = \infty$ and $l_k = 0$), and for the particular case of having equality mutual capacity constraints. Therefore, the heuristic of Figure 3 can be used for three-dimensional tables replacing line 6 by the solution of a multicommodity network flows problem.

5.2   Linked and Hierarchical Tables

Linked tables can be defined as tables that share some common cells or, more generally, whose entries are linked by some linear relation. Hierarchical tables (tables with subtotals) can be considered a particular case of linked tables, in which the common cells correspond to subtotal entries. Standard network flows models are only useful for hierarchical tables in one dimension, as shown in [18]. We focus on the general linked tables model.

Given a linked table made of t two or three-dimensional tables, and a set of four-dimensional elements with the information of the common cells,

$$\mathcal{E} = \{(u, r, v, s) : \text{cell } u \text{ in table } r \text{ must be equal to cell } v \text{ in table } s\},$$

the overall table relations can be written as

$$A^i a^i = b^i, \quad i = 1 \ldots t \tag{13}$$
$$a_u^r - a_v^s = 0, \quad (u, r, v, s) \in \mathcal{E}. \tag{14}$$

$A^i a^i = b^i$ denote the network or multicommodity network equations, depending of the dimension of table i, while (14) impose the same value for the common cells. Clearly, (14) and the mutual capacity equations (12) of all the three-dimensional tables form a set of linear side constraints, that, together with the remaining network equations, match the model defined in (7). Therefore, linked (and hierarchical) tables can be modeled as a network with side constraints.

6   Preliminary Computational Results

We implemented an extension for three-dimensional tables of the heuristic described in Section 3.3, including a generalization of the lower-bounding procedure introduced in [19]. It was coded in AMPL [12]. Since no specialized multicommodity solver is currently linked to AMPL, we used the network+dual option of the Cplex 6.5 package [17]. This option finds first a feasible point for the network constraints using a specialized network primal simplex algorithm; this point is then used as a warm start for the dual simplex. The two generators of Section 4 were extended for three-dimensional tables. Table 1 reports the dimensions and results of the instances. Column "gen" shows the generator used. Column "$m \times n \times l$" gives the size of the table. Column "p" is the number of primary suppressions. Columns "c.s." and "v.s." show respectively the total number of suppressed cells and of value suppressed (including primary suppressions). Column "nnf" is the number of multicommodity network flows problems solved, while column "CPU_nf" gives the overall CPU time spent in their solution. For all the instances the upper and lower protection levels were a 15% of the cell value. The execution environment was the same that for Section 4. The figures of the results columns (last four) are significantly greater than those obtained for two-dimensional tables with a similar number of cells and primary suppressions. This weakness of the heuristic is due to the complex cell interrelations of three-dimensional tables.

Table 1. Dimensions and results for 3D tables (using network+dual solver)

gen.	$m \times n \times l$	p	c.s.	v.s.	nnf	CPU_nf
1	$10 \times 10 \times 10$	50	254	67082	76	18.7
1	$10 \times 10 \times 10$	100	234	75317	110	31.3
1	$10 \times 10 \times 20$	50	317	77439	81	31.3
1	$10 \times 10 \times 20$	100	362	94765	127	82.9
1	$10 \times 20 \times 10$	50	298	79644	69	29.2
1	$10 \times 20 \times 10$	100	397	97964	115	102.6
1	$10 \times 20 \times 20$	50	473	97753	86	127.1
1	$10 \times 20 \times 20$	100	526	118745	144	224.2
2	$10 \times 10 \times 10$	50	170	10099	43	7.8
2	$10 \times 10 \times 10$	100	191	6194	38	9.1
2	$10 \times 10 \times 20$	50	222	14458	65	16.6
2	$10 \times 10 \times 20$	100	275	12027	43	24.6
2	$10 \times 20 \times 10$	50	222	14192	63	21.7
2	$10 \times 20 \times 10$	100	310	16035	89	40.0
2	$10 \times 20 \times 20$	50	296	19252	75	54.5
2	$10 \times 20 \times 20$	100	398	19841	103	86.9

We also solved the set of instances using the dual simplex option of Cplex 6.5, which does not exploit the network structure of the problem. The results are shown in Table 2. Clearly, the execution times drastically reduced in most instances. This is a surprising result, specially because the network+dual Cplex option is a highly regarded algorithm for multicommodity flows and flows with side constraints. A possible explanation is that the problem solved by the network primal simplex method is very degenerate (i.e., many basic variables at bounds) which means a large number of unproductive iterations. Additional experiments with larger instances and alternative network flow solvers have to be performed before concluding that the dual simplex method is the most efficient approach for three-dimensional tables.

Table 2. Dimensions and results for 3D tables (using dual solver)

gen.	$m \times n \times l$	p	c.s.	v.s.	nnf	CPU_nf
1	$10 \times 10 \times 10$	50	227	60961	56	7.6
1	$10 \times 10 \times 10$	100	259	78641	69	10.1
1	$10 \times 10 \times 20$	50	307	74239	74	19.0
1	$10 \times 10 \times 20$	100	384	97304	79	26.8
1	$10 \times 20 \times 10$	50	292	74327	60	15.5
1	$10 \times 20 \times 10$	100	446	103971	82	33.4
1	$10 \times 20 \times 20$	50	480	95978	73	64.1
1	$10 \times 20 \times 20$	100	608	124817	104	96.4
2	$10 \times 10 \times 10$	50	170	10099	38	4.6
2	$10 \times 10 \times 10$	100	190	6115	31	4.0
2	$10 \times 10 \times 20$	50	222	14458	65	15.5
2	$10 \times 10 \times 20$	100	261	10889	24	7.5
2	$10 \times 20 \times 10$	50	222	14192	63	14.7
2	$10 \times 20 \times 10$	100	306	15656	81	19.7
2	$10 \times 20 \times 20$	50	296	19252	75	50.5
2	$10 \times 20 \times 20$	100	390	19558	93	50.4

7 Conclusions

From our computational experience, it can be concluded that the efficiency of
the two heuristics evaluated for two-dimensional tables depends on the particular structure of the instances to be solved. We also reported some preliminary
results with a network flows heuristic for three-dimensional tables. Among the
future tasks to be done we find a comprehensive evaluation of multicommodity
models for three-dimensional tables, including larger instances and alternative
specialized solvers, and the computational study of a heuristic for linked and
hierarchical tables based on network flows with side constraints.

References

1. Ahuja, R.K, Magnanti, T.L., Orlin, J.B.: Network Flows. Prentice Hall (1993)
2. Birge, J.R., Louveaux, F.: Introduction to Stochastic Programming. Springer
   (1997)
3. Bixby, R.E., Fenelon, M., Gu, Z., Rothberg, E., Wunderling, R.: MIP: Theory
   and practice—Closing the gap. In System Modelling and Optimization. Methods,
   Theory and Applications, eds. M.J.D. Powell and S. Scholtes. Kluwer (2000) 19–49
4. Carvalho, F.D., Dellaert, N.P., Osório, M.D.: Statistical disclosure in two-
   dimensional tables: general tables. J. Am. Stat. Assoc. **89** (1994) 1547–1557
5. Castro, J.: A specialized interior-point algorithm for multicommodity network
   flows. SIAM J. on Opt. **10** (2000) 852–877
6. Castro, J., Nabona, N. An implementation of linear and nonlinear multicommodity
   network flows. European Journal of Operational Research **92** (1996) 37–53
7. Cox, L.H.: Network models for complementary cell suppression. J. Am. Stat. Assoc.
   90 (1995) 1453–1462

8. Dellaert, N.P., Luijten, W.A.: Statistical disclosure in general three-dimensional tables. Statistica Neerlandica **53** (1999) 197–221

9. Fischetti, M., Salazar, J.J.: Models and algorithms for the 2-dimensional cell suppression problem in statistical disclosure control. Math. Prog. **84** (1999) 283–312

10. Fischetti, M., Salazar, J.J.: Models and algorithms for optimizing cell suppression in tabular data with linear constraints. J. Am. Stat. Assoc. **95** (2000) 916–928

11. Fischetti, M., Salazar, J.J.: Partial cell suppression: a new methodology for statistical disclosure control. Working paper, Department of Statistics, Operations Research and Computing, University of La Laguna (1998)

12. Fourer, R, Gay, D.M., Kernighan, B.W.: AMPL: A Modeling Language for Mathematical Programming. Duxbury Press (1993)

13. Frangioni, A., Gallo, G: A bundle type dual-ascent approach to linear multicommodity min cost flow problems. INFORMS J. on Comp. **11** (1999) 370–393

14. Giessing, S.: New tools for cell-suppression in τ-Argus: one piece of the CASC project work draft. Joint ECE/Eurostat Work Session on Statistical Data Confidentiality, Skopje (2001).

15. Goldberg, A.V., Oldham, J.D., Plotkin, S., Stein, C.: An implementation of a combinatorial approximation algorithm for minimum-cost multicommodity flow. In Lecture Notes in Computer Sciences. Proceedings of the 6th International Integer Programming and Combinatorial Optimization Conference, eds. R.E. Bixby, E.A. Boyd and R.Z. Ríos-Mercado. Springer (1998)

16. Gondzio, J, Sarkissian, R.: Parallel interior-point solver for structured linear programs. Technical Report MS-00-025, Department of Mathematics and Statistics, The University of Edinburgh (2000)

17. ILOG CPLEX: ILOG CPLEX 6.5 Reference Manual Library. ILOG (1999)

18. Jewett, R.: Disclosure analysis for the 1992 Economic Census. Manuscript, Economic Programming Division, Bureau of the Census (1993)

19. Kelly, J.P., Golden, B.L, Assad, A.A.: Cell Suppression: disclosure protection for sensitive tabular data. Networks **22** (1992) 28–55

20. Massell, P.B.: Cell suppression and audit programs used for economic magnitude data. SRD Research Report Series RR2001/01 (2001).

21. Robertson, D.: Improving Statistic's Canada cell suppression software (CONFID). Proceedings of Compstat 2000.

HiTaS: A Heuristic Approach to Cell Suppression in Hierarchical Tables

Peter-Paul de Wolf*

Statistics Netherlands (CBS)
P.O. Box 4000, 2270 JM Voorburg, The Netherlands
pwof@cbs.nl

Abstract. This paper describes a heuristic approach to find suppression patterns in tables that exhibit a hierarchical structure in at least one of the explanatory variables. The hierarchical structure implies that there exist (many) sub-totals, i.e., that (many) sub-tables can be constructed. These sub-tables should be protected in such a way that they cannot be used to undo the protection of any of the other tables. The proposed heuristic approach has a top-down structure: when a table of high level (sub-)totals is suppressed, its interior turns into the marginals of possibly several tables on a lower level. These lower level tables are then protected while keeping the marginals fixed.

Keywords: Statistical Disclosure Control, table suppression, hierarchical structure, top-down method.

1 Introduction

In statistical disclosure control (SDC), it is common practice to protect data published in tables, one table at a time. At Statistics Netherlands a software package called τ-ARGUS is developed to facilitate several SDC methods to protect tabular data. However, in practice different tables are often linked to each other, in the sense that certain cells can occur in these tables simultaneously. E.g., marginal cells in one table might well be the same cells that appear in the interior of another table. When dealing with such sets of tables, the used disclosure control methods should be consistent with each other: it should not be possible to undo the protection of a table using another – by itself safe – table.

In this paper, the disclosure control method of cell suppression is considered. Based on a sensitivity measure (e.g., a dominance rule and/or a minimal frequency rule (see, e.g., [1])) it is decided which cells of a certain table need to be suppressed. These suppressions are called primary suppressions. However, in order to eliminate the possibility to recalculate these suppressed cells (either exactly or up to a good approximation), additional cells need to be suppressed, which are called secondary suppressions. Whenever a variable has a hierarchical structure (e.g., regional variables, classification variables like NACE), secondary suppressions might imply even more secondary suppressions in related tables. In

---

* The views expressed in this paper are those of the author and do not necessarily reflect the policies of Statistics Netherlands.

J. Domingo-Ferrer (Ed.): Inference Control in Statistical Databases, LNCS 2316, pp. 74–82, 2002.

[2] a theoretical framework is presented that should be able to deal with hierarchical and generally linked tables. In that frame-work additional constraints to a linear programming problem are generated. The number of added constraints however, grows rapidly when dealing with hierarchical tables, since many dependencies exist between all possible (sub-)tables containing many (sub-)totals. A heuristic approach will be proposed that deals with a large set of (sub-)tables in a particular order. In the next section the approach will be discussed in more detail and a simple example with artificial data will be used to illustrate the ideas. Section 3 presents results concerning the example introduced in Section 2. In Section 4 some additional remarks are made on the presented cell-suppression method. Appendix A contains all the sub-tables defined in the example of Section 2 that have to be considered when applying the method.

2   Hierarchical Cell Suppression

The proposed heuristic approach in constructing a method to deal with cell suppression in hierarchical tables is a top-down approach. For the purpose of simplicity of the exposition, we will discuss the method in case of a two dimensional table in which both explanatory variables exhibit a hierarchical structure. The ideas are easily extended to the situation of higher dimensional tables, with an arbitrary number of hierarchical explanatory variables. Similar ideas (partitioning the complex table into several less complex subtables) have been discussed in [3].

The example that will be used to illustrate the ideas consists of two fictitious hierarchical variables: a regional variable R (Region) and a classifying variable BC (Industrial Classification). The hierarchical structures are represented in Figures 1 and 2. The variable R consists of 4 levels: level 0 (R), level 1 ($P1$, $P2$, $P3$), level 2 ($C21$, $C22$, $C31$, $C32$) and level 3 ($D211$, $D212$). The variable BC consists of 3 levels: level 0 (BC), level 1 (I, A, O) and level 2 (LI, MI, SI, LA, SA).

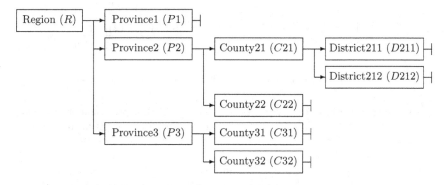

Fig. 1. Hierarchical structure of regional variable R

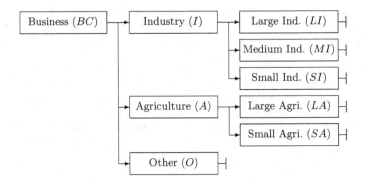

Fig. 2. Hierarchical structure of classifying variable BC

The first step is to determine the primary unsafe cells in the base-table consisting of all the cells that appear when crossing the two hierarchical variables. This way all cells, representing a (sub-)total or not, are checked for primary suppression. The base-table of the example is represented in Figure 3, in which the cells that primarily will be suppressed are marked by a boldface value and an empty cell is denoted by a minus-sign $(-)$.

				BC								
					I				A			O
						LI	MI	SI		LA	SA	
R				300	125	41	44	40	93	51	42	82
	$P1$			77	31	8	12	11	26	13	13	20
	$P2$			128	52	21	18	13	44	24	20	32
		$C21$		77	25	11	9	**(5)**	31	18	13	21
			$D211$	35	16	4	7	**(5)**	10	6	**(4)**	**(9)**
			$D212$	42	9	7	**(2)**	–	21	12	9	12
		$C22$		51	27	10	**(9)**	8	13	6	7	11
	$P3$			95	42	12	14	16	23	14	9	30
		$C31$		45	27	8	9	10	**(5)**	**(2)**	**(3)**	13
		$C32$		50	15	4	5	6	18	12	6	17

Fig. 3. Base-table of $R \times BC$, (**bold**) means primary unsafe

Knowing all primary unsafe cells, the secondary cell suppressions have to be found in such a way, that each (sub-)table of the base-table is protected and that the different tables cannot be combined to undo the protection of any of the other (sub-)tables.

There are (at least) two possible approaches to this problem: a bottom-up and a top-down approach. Each approach attempts to protect a large set of (sub-)tables in a sequential way. In the bottom-up approach one would move upwards in the hierarchy of the explanatory variables and calculate the secondary suppressions using the (fixed) suppression pattern of a lower level table as its interior. In the example, one of the lowest level tables to consider is $(D211, D212) \times (LA, SA)$. Since the cell $(D211, SA)$ is primarily unsafe, some additional suppressions have to be found. One possible suppression pattern is given by suppressing the interior, while keeping the marginals publishable. See Figure 4.

	LA	SA	A
D211	6	(4)	10
D212	12	9	21
C21	18	13	31

$\implies$

	LA	SA	A
D211	[6]	(4)	10
D212	[12]	[9]	21
C21	18	13	31

Fig. 4. Low level table, **(bold)** is primary, [normal] is secondary suppression

However, the higher level table $(D211, D212) \times (I, A, O)$ has a primary unsafe cell $(D211, O)$ and needs secondary suppressions as well. This could lead to a (secondary) suppression of cell $(D211, A)$. See Figure 5.

	I	A	O	BC
D211	16	10	(9)	35
D212	9	21	12	42
C21	25	31	21	77

$\implies$

	I	A	O	BC
D211	16	[10]	(9)	35
D212	9	[21]	[12]	42
C21	25	31	21	77

Fig. 5. Higher level table, **(bold)** is primary, [normal] is secondary suppression

Unfortunately, one of the secondarily suppressed cells is also a marginal cell of the previously considered table, hence backtracking is needed: the table of Figure 4 has to be considered again, using the information of the suppression pattern of the table in Figure 5. In other words, the tables can not be dealt with independently of each other.

The basic idea behind the top-down approach is to start with the highest levels of the variables and calculate the secondary suppressions for the resulting table. In theory the first table to protect is thus given by a crossing of level 0 of variable R with level 0 of variable BC, i.e., the grand total. The interior of the protected table is then transported to the marginals of the tables that appear when crossing lower levels of the two variables. These marginals are then 'fixed' in the calculation of the secondary suppressions of that lower level table, i.e., they are not allowed to be (secondarily) suppressed. This procedure is then repeated until the tables that are constructed by crossing the lowest levels of the two variables are dealt with.

Using the top-down approach, the problems that were stated for the bottom-up approach are circumvented: a suppression pattern at a higher level only introduces restrictions in the marginals of lower level tables. Calculating secondary suppressions in the interior while keeping the marginals fixed, is then independent between the tables on that lower level, i.e., no backtracking is needed.

Moreover, added primary suppressions in the interior of a lower level table are dealt with at that same level: secondary suppressions can only occur in the same interior, since the marginals are kept fixed.

However, when several empty cells are apparent in a low level table, it might be the case that no solution can be found if one is restricted to suppress interior cells only. Unfortunately, backtracking is then needed as well. However, this is usually restricted to one higher level, whereas in the case of the bottom-up approach one could have to deal with all lower level (sub-)tables.

The introduction of suppressions in the marginals of a table requires additional (secondary) suppressions in its interior. Most models that are used to find suppression patterns can only deal with either safe cells or primary unsafe cells. Applying such models in the present situation implies that secondary suppressions temporarily need to be considered as primary unsafe cells if they occur in other (linked) sub-tables. Usually, rather large safety-ranges (see e.g., [1]) are imposed on primary unsafe cells, which may be too restrictive in the case of secondary suppressions. Actually, the suppression pattern of the higher level table that produced the secondary suppression for the marginal of a lower level table, defined the amount of safety-range that should be put on the secondary suppression. Unfortunately, in most cases the suppression pattern is such that it is virtually impossible to determine which secondary suppressions are used for which primary suppressions. Hence, the needed safety-range for the secondary suppression is not easily available. In practice it might be more convenient to manually impose rather small safety-ranges on these cells, in order to eliminate the possibility that the value can be recalculated exactly, while being less restrictive as in the case of 'real' primary unsafe cells. Moreover, one could take the maximum safety-range of all the primary suppressions as an upper bound for the safety-range of any secondary suppression of that (sub-)table.

Obviously, all possible (sub)tables should be dealt with in a particular order, such that the marginals of the table under consideration have been protected as the interior of a previously considered table. This can be assured by defining certain classes of tables. First define a group by a crossing of levels of the explanatory variables. The classes of tables mentioned before then consist of the groups in which the numbers of the levels add to a constant value. Table 1 contains all the classes and groups that can be defined in the example.

Table 1. The classes defined by crossing R with BC

Class	Groups
0	00
1	10, 01
2	20, 02, 11
3	21, 12, 30
4	22, 31
5	32

E.g., group 31 is a crossing of level 3 of variable R and level 1 of variable BC. Such a group thus may consist of several tables: group 31 consists of the table $(D211, D212) \times (I, A, O)$ whereas group 21 consists of the tables $(C21, C22) \times (I, A, O)$ and $(C31, C32) \times (I, A, O)$. Note that the latter two tables need to be dealt with separately. Class 4 consists of group 22 $(2 + 2 = 4)$ and group 31 $(3 + 1 = 4)$.

Defined in this way, marginals of the tables in class i have been dealt with as the interior of tables in a class j with $j < i$. As a result, each table in class i can be protected independently of the other tables in that particular class, whenever the tables in classes j with $j < i$ have been dealt with.

The number of tables in a group is determined by the number of parent-categories the variables have one level up in the hierarchy. A parent-category is defined as a category that has one or more sub-categories. E.g., group 22 has four tables, since variable R has two parent-categories at level 1 (categories $P2$ and $P3$) and variable BC has two parent-categories at level 1 (categories I and A) and thus $2 \times 2 = 4$ tables can be constructed.

To illustrate things even further, in Appendix A all tables are given that had to be checked in case of the example, ordered by the corresponding classes and groups as given in Table 1.

3 Examples of Output

A prototype C++ 32-bits Windows console program HiTaS has been written, to perform the proposed heuristic approach to cell-suppression in hierarchical tables. Figure 6 contains the results of running that program on the example of Section 2. The unprotected table was given in Figure 3. In Figure 6 primary suppressions are represented by the layout **(bold)**, the secondary suppressions by [normal].

			BC	*I*				*A*			*O*
					LI	*MI*	*SI*		*LA*	*SA*	
R			300	125	41	44	40	93	51	42	82
	P1		77	31	8	12	11	26	13	13	20
	P2		128	52	21	18	13	44	24	20	32
		C21	77	25	[11]	[9]	**(5)**	31	18	13	21
		D211	35	[16]	[4]	[7]	**(5)**	10	[6]	**(4)**	**(9)**
		D212	42	[9]	[7]	**(2)**	–	21	[12]	[9]	[12]
		C22	51	27	[10]	**(9)**	[8]	13	6	7	11
	P3		95	42	12	14	16	23	14	9	30
		C31	45	[27]	[8]	[9]	[10]	**(5)**	**(2)**	**(3)**	13
		C32	50	[15]	[4]	[5]	[6]	[18]	[12]	[6]	17

Fig. 6. Base-table of $R \times BC$ after applying HiTaS, **(bold)** is primary and [normal] is secondary suppression

To illustrate the size of the problem and the associated running time to complete the cell-suppression of a hierarchical table, consider the following more realistic and much more complex example.

Consider a data file with 7856 records containing four variables: three hierarchical explanatory variables and one response variable. The explanatory variables are:

- A regional variable (17 codes, including the (sub-)totals) with a hierarchical structure of 3 levels (level 0 up to level 2)
- A business classification variable (54 codes, including the (sub-)totals) with a hierarchical structure of 6 levels (level 0 up to level 5)
- A business-size variable (19 codes, including the (sub-)totals) with a hierarchical structure of 5 levels (level 0 up to level 4)

This resulted in a base-table with 20672 cells of which 11911 were empty and 4267 were primary unsafe. The number of (sub-)tables to be considered was 13097. The presence of many empty cells resulted in the fact that 36 times a backtracking was needed. HiTaS concluded that 2191 secondary suppressions were needed to completely protect the hierarchical table consistently.

To complete all this, the program needed about 47 minutes on an Intel Celeron 200 MHz Windows95 PC with 128 Mb RAM.

4   Remarks and Future Developments

In the present paper, a heuristic approach is given to find a suppression pattern in a hierarchical table. Whether or not the resulting pattern is acceptable still needs further evaluation. Moreover, it would be interesting to compare the results with results that would be generated by more sophisticated methods like the one proposed in [2].

Another interesting research topic would be to investigate to what extent the proposed heuristic approach for hierarchical tables can be speeded up by considering a certain subset of all the (sub-)tables.

Moreover, the partitioning of a hierarchical table into classes of smaller (sub-)tables could be reconsidered. The partitioning of the hierarchical table into classes consisting solely of non-hierarchical tables, implies that a specific problem, discussed in e.g., [4], might still be present. I.e., the proposed approach is neither completely safe nor optimal in the sense described in [2]. It might be interesting to investigate whether this could be solved considering different partitions of the hierarchical base-table.

The actual implementation of the method includes some interesting future developments as well. First of all, this approach will be incorporated into the more user friendly software package τ-ARGUS in the fifth framework project CASC. Moreover, it could be interesting to investigate the effect of parallel computing on the overall running time. Since the structure of the method is such that several tables can be dealt with simultaneously, it seems rather straightforward to implement parallel computing.

References

1. Willenborg, L. and T. de Waal (1996). *Statistical Disclosure Control in practice.* Lecture Notes in Statistics 111, Springer-Verlag, New York.
2. Fischetti, M. and J.J. Salazar-González (1998). Models and Algorithms for Optimizing Cell Suppression in Tabular Data with Linear Constraints. Technical Paper, University of La Laguna, Tenerife.
3. Cox, L.H. (1980). 'Suppression Methodology and Statistical Disclosure Control', *Journal of the American Statistical Association,* volume 75, number 370, pp. 377–385.
4. Sande, G. (2000). Blunders by official statistical agencies while protecting the confidentiality of business statistics, submitted to IS Review.

Appendix A: Tables to Be Checked in Example of Section

2 This appendix contains all tables that had to be checked for secondary suppression in case of the example of Section 2. The base-table (only needed to determine the primary cell-suppressions) is shown in Figure 3 of Section 2 itself. The group names and class numbers are explained in Section 2. Note that the table of group 00 is already dealt with in the base-table.

In each table, primary suppressions are marked **(bold)** and secondary suppressions are marked [normal].

Class 0
Group 00

	BC
R	300

Class 1
Group 10   *Group 01*

	BC
P1	77
P2	128
P3	95
R	300

	I	A	O	BC
R	125	93	82	300

Class 2
Group 20

	BC
C21	77
C22	51
P2	128

	BC
C31	45
C32	50
P3	95

Group 02

	LI	MI	SI	I
R	41	44	40	125

	LA	SA	A
R	51	42	93

Group 11

	I	A	O	BC
P1	31	26	20	77
P2	52	44	32	128
P3	42	23	30	95
R	125	93	82	300

Class 3

Group 21

	I	A	O	BC
C21	25	31	21	77
C22	27	13	11	51
P2	52	44	32	128

	I	A	O	BC
C31	[27]	(5)	13	45
C32	[15]	[18]	17	50
P3	42	23	30	95

Group 12

	LI	MI	SI	I
P1	8	12	11	31
P2	21	18	13	52
P3	12	14	16	42
R	41	44	40	125

	LA	SA	A
P1	13	13	26
P2	24	20	44
P3	14	9	23
R	51	42	93

Group 30

	BC
D211	35
D212	42
C21	77

Class 4

Group 22

	LI	MI	SI	I
C21	[11]	[9]	(5)	25
C22	[10]	(9)	[8]	27
P2	21	18	13	52

	LA	SA	A
C21	18	13	31
C22	6	7	13
P2	24	20	44

	LI	MI	SI	I
C31	[8]	[9]	[10]	[27]
C32	[4]	[5]	[6]	[15]
P3	12	14	16	42

	LA	SA	A
C31	(2)	(3)	(5)
C32	[12]	[6]	[18]
P3	14	9	23

Group 31

	I	A	O	BC
D211	[16]	10	(9)	35
D212	[9]	21	[12]	42
C21	25	31	21	77

Class 5

Group 32

	LI	MI	SI	I
D211	[4]	[7]	(5)	16
D212	[7]	(2)	–	9
C21	[11]	[9]	(5)	25

	LA	SA	A
D211	[6]	(4)	10
D212	[12]	[9]	21
C21	18	13	31

Model Based Disclosure Protection

Silvia Polettini[1], Luisa Franconi[1], and Julian Stander[2]

[1] ISTAT, Servizio della Metodologia di Base per la Produzione Statistica
Via Cesare Balbo, 16, 00185 Roma, Italy
{polettin,franconi}@istat.it
[2] University of Plymouth,School of Mathematics and Statistics
Drake Circus, PL4 8AA Plymouth, UK
J.Stander@plymouth.ac.uk

Abstract. We argue that any microdata protection strategy is based on a formal reference model. The extent of model specification yields "parametric", "semiparametric", or "nonparametric" strategies. Following this classification, a parametric probability model, such as a normal regression model, or a multivariate distribution for simulation can be specified. Matrix masking (Cox [2]), covering local suppression, coarsening, microaggregation (Domingo-Ferrer [8]), noise injection, perturbation (e.g. Kim [15]; Fuller [12]), provides examples of the second and third class of models. Finally, a nonparametric approach, e.g. use of bootstrap procedures for generating synthetic microdata (e.g. Dandekar *et. al.* [4]) can be adopted.

In this paper we discuss the application of a regression based imputation procedure for business microdata to the Italian sample from the Community Innovation Survey. A set of regressions (Franconi and Stander [11]) is used for generating flexible perturbation, for the protection varies according to identifiability of the enterprise; a spatial aggregation strategy is also proposed, based on principal components analysis. The inferential usefulness of the released data and the protection achieved by the strategy are evaluated.

Keywords: Business microdata, confidentiality, performance assessment, protection models, regression models.

1 Introduction

Dissemination of statistical information while protecting confidentiality of respondents has always been the mission of National Statistical Institutes (NSIs). In the past this was achieved by producing aggregated data.

Dissemination of microdata that allow for reanalysis by different users with different aims is the challenge that NSIs have been facing in the last few years.

The protection of confidentiality of respondents is, intuitively, the protection of their identity. A disclosure occurs when an association is made between a respondent and a record in the released microdata file. We call such association an *identification*.

J. Domingo-Ferrer (Ed.): Inference Control in Statistical Databases, LNCS 2316, pp. 83–96, 2002.
© Springer-Verlag Berlin Heidelberg 2002

The identification of a respondent allows for the disclosure of all the confidential information provided. Identification may be possible using *a priori* information and other data generally available from public registers.

Disclosure limitation strategies aim at reducing the information content of the released microdata in order to make identification a difficult and uncertain task. In this paper we argue that any microdata protection strategy is based on a formal reference model. We use this view to show that different disclosure limitation methods present in the literature can be seen in a unified manner, distinguishing them according to the degree and number of restrictions imposed on the model. This is discussed in Section 2.

In Section 3 we present the results of an extended study of the application of the model proposed by Franconi and Stander in [11] on the Italian sample of the Community Innovation Survey (CIS). Our experience with CIS data reveals several issues that need to be addressed, such as the use of robust methods, the accurate choice of some parameters, the diagnosis of the protection model, the usefulness of the data and so on. These will be discussed in Section 3.

Section 4 contains conclusions and suggestions for further research.

2   A Unified Framework for Model Based Protection

In this section we express our view about protection methods for data confidentiality. We present a unified framework in which each protection method has its own reference *model*, at least in a broad sense.

In our view, in order to release protected data the NSIs have basically two options:

1. coarsening, e.g. transforming the data (rows or columns of the data matrix). An extreme version of this consists of artificially introducing missing values (*local suppression*), which includes subsampling as a special case;
2. simulating artificial data set(s) or records.

Coarsening consists of transforming the data by using deterministic or random functions, either invertible or singular. Little [16] suggests releasing a summary of the data themselves, such as a set of sufficient statistics for the assumed model. An aggregated (marginal) table for categorical data is an example of this. This is also an example of a non invertible transformation -unless the sufficient statistic achieves no reduction of the data. At the extreme of such an approach is reducing the sensitive information carried by the data by artificially introducing missing values. In both cases, Little [16] discusses post-release analysis of protected data by means of exact or approximate ("pseudo") likelihood inference, heavily relying on the EM algorithm (see [7], [17]).

Full imputation, i.e. generation of a set of artificial units, is an alternative option for NSIs. The idea of releasing a synthetic sample seems a way to avoid any confidentiality threat, as the confidentiality of synthetic individuals is not of concern. Rubin [19] proposes using multiply-imputed data sets for release, and states the superiority of this approach over other methods of data protection. Difficulties in application of these ideas are documented by Kennickel [14].

The idea of simulation is connected with the principle that the statistical content of the data lies in the likelihood, not in the information provided by the single respondents. Consequently, a model well representing the data could in principle replace the data themselves; alternatively, a simulated data set (or a sample of data sets) drawn from the above mentioned model can represent a more user-friendly release strategy. Indeed, since the proposal of Rubin [19], several authors have stated the usefulness of disseminating synthetic data sets; see Fienberg et al. [9], and, more recently, Grim et al. [13] and Dandekar et al. [4], [5]. In the choice of the generating model, the aim is always to reproduce some key characteristics of the sample.

Each of the alternative options discussed so far can be considered an imputation procedure: a protection model is formalised and released values are generated according to it in substitution of the original ones.

In the sequel, we will consider an observed data matrix X of dimension n by k; the columns of X will correspond to variables and will be denoted by X_l, $l = 1, \ldots, k$. A tilde will denote the corresponding released quantities, so that $\tilde{X}$ will denote the released matrix, $\tilde{X}_l$ the released l-th variable.

We suggest that the basic ingredient of any technique of statistical disclosure control is the *protection model*. As illustrated by the examples provided above, by protection model we mean a relation expressing the protected data in terms of the original data by means of some transformation function. Using this formulation, for the released data $\tilde{X}$ a class of distributions can be specified either directly, or through assumptions about the family of laws governing the matrix X. The degree of specification of the distributional component in the protection model varies from model to model. Some methods make no distributional assumptions, some others specify a parametric class for the probability law of the released data through assumptions on the original data matrix. Moreover, sometimes only a component of the protection model is given a fixed distribution. In some cases, only some characteristics of the distributions, such as conditional means, are specified. In this sense we will distinguish between fully nonparametric, semiparametric and fully parametric models, and from these devise nonparametric, semiparametric and fully parametric methods for data confidentiality. In this view, it is the extent of formalisation of the model which makes the strategies inherently different.

2.1   Nonparametric Protection Methods

Suppose that the distribution of $\tilde{X}$ is left completely unspecified. Suppose further that the protection model for the released matrix $\tilde{X}$ has the form of a matrix mask, $\tilde{X} = XB$. The last expression is a compact notation encompassing several different imputation procedures, as discussed in Little [16] and formalised in Cox [2]. As the latter author has shown, this protection model may provide locally suppressed, microaggregated or swapped data depending on the choice of the *attribute transforming mask B*.

Use of an additive component in the mask accounts for other types of transformations, such as topcoding. In this case, the model takes the more general form $\tilde{X} = XB + C$.

Exclusion of selected units is accomplished by using a different matrix mask, acting on the rows: $\tilde{X} = AX$, A being termed a *record transforming mask*.

Finally, exclusion of selected units followed by deletion of some pre-specified attributes is accomplished by the more general matrix mask $\tilde{X} = AXB$; actually Cox [2] uses the more general notation $X = AXB + C$.

For protection by simulating artificial data sets, the use of procedures such as the bootstrap, or modified versions of it, give rise to nonparametric disclosure limitation strategies. An example of this is the proposal in [4], [5] based on Latin Hypercube Sampling, in which the empirical cumulative distribution function is used to create equally probable intervals which allow to draw stratified samples. The work by Fienberg *et al.* [9] discusses analogous strategies that we would classify as nonparametric protection methods.

2.2   Semiparametric Protection Methods

In the previous paragraph, the model contains nothing but the empirical distribution of the data, plus known constants. A semiparametric structure is introduced through assumptions about the masking matrices A, B and C and/or the observed matrix X.

In particular, let us introduce a random matrix C having a known distribution. Then the masked matrix $\tilde{X}$ obtained by adding to AXB a realization of C represents a perturbation of the original data.

Of course, C could be masked by introducing a column-selecting matrix D to be used whenever a variable need not be noise injected. In the context of database protection, Duncan and Mukherjee [6] analyse the extent to which the perturbation method can be applied if valid (e.g. precise) inferences about parameters of interest are to be drawn. In particular, they discuss bounds on the variance of the noise distribution.

A particular case of semiparametric masking is the model discussed in [16], which replaces the observed data by the sample mean plus random noise, obtained by setting $A = \mathbf{1}_{n \times k}$. The model just discussed in general prescribes for the data to be released a convolution of the distribution of the data, possibly after suitable transformation, with the noise distribution.

For a thorough, up-to-date discussion of noise injection, refer to the paper by Brand in this volume [1].

We also define semiparametric the imputation model based on least squares regression; it is a semiparametric version of the naive strategy based on the release of sample mean. The model prescribes a relation between a variable X_l to be protected and a set of covariates extracted from the observed matrix X, that we will denote by $X_{K \setminus l}$, $K \subseteq \{1, 2 \dots, k\}$, without further assumptions on the error distribution. For a similar argument, see [16].

2.3   Parametric Models

A step further is represented by the specification of a class of distributions for the released data. If the variables are continuous, very often the multivariate normal distribution is used, possibly after a normalising transformation.

One option in model based protection is disseminating the fitted values of a normal regression model for one or more variables X_l in X; this is the main idea in [11]. A slight variation (see [16]) of the regression method consists of releasing the predicted value plus random noise, taken from the estimated error distribution. This aims to compensate for the reduction of variability of the fitted values compared to that of the original data.

Of course the protection strategy based on regression models may be confined to some of the units of the data matrix; in the previous notation, this may be represented symbolically as $\tilde{X}_l = A(X_{K \setminus l}\hat{\beta} + \eta)$, $\eta \sim N(0, \hat{\sigma}^2_{X_l | X_{K \setminus l}})$.

Another example of parametric disclosure protection is the release of prediction intervals for the variables to be protected, based on distributional assumptions, with or without a regression model for the variables to be protected (for an analogous strategy, see [10]).

Finally, simulation of artificial data sets can be based on a fully specified model; the mixture model adopted in [13] and estimated by likelihood methods with the aid of the EM algorithm provides an example of parametric protection.

For categorical variables, the strategy of releasing synthetic data sets drawn from a nonsaturated loglinear model "capturing the essential features of the data", as proposed in [9], is another example of parametric procedure.

Several proposals in the literature are present which take advantage of a Bayesian formulation: among the others, [10] develop a Bayesian hierarchical model with spatial structure, making use of the MCMC output to release predictive intervals instead of single values.

For a review of the Bayesian approach to data disclosure, see Cox [3].

3   A Regression Model Approach for Protection of Business Microdata: Application to the CIS Survey

3.1   The Microdata

The microdata used is the Italian sample from the CIS. This is a survey of technological innovation in European manufacturing and services sector enterprises. The variables present in the survey can be classified into two sets according to whether or not they allow an identification. The first set contains all the general information about the enterprise such as its main economic activity (four digit NACE rev. 1 Classification), geographical area (eight regions based on NUTS1), number of employees (integer ≥ 20), turnover, exports and total expenses for research and innovation (for year 1996, measured in millions of Italian lire). These variables either appear in public registers or, being quantitative, can reveal the size of an enterprise. This information on the size, when used with the others, can be extremely dangerous as it provides clues about the largest and therefore most identifiable enterprises.

We assume that the more the enterprise is outlying, the higher the risk of it being identified. For an assessment of the protection achieved by the method we also rely on outliers (see Section 3.3). This approach has a limit in its qualitative nature; in fact, a quantitative measure of disclosure based on record linkage as in [20] could be pursued.

The second set of variables contains confidential information on a range of issues connected with innovation. Because of their nature, they do not allow an identification of the enterprise.

3.2   The Protection Model

In this section we summarise the method proposed by Franconi and Stander in [11]. The underlying idea is to make identification a difficult task by reducing the information content of all the identifying variables. For the CIS data such variables are: economic activity, geographical area, number of employees, turnover, exports and total expenditure for research and innovation. For the variables main economic activity and geographical area the release of broader categories is proposed. Economic activity is aggregated according to the first two digit of the NACE rev. 1 classification; geographical area is given two broader categories based on principal component analysis; for further details see [11]. To reduce the information content of the quantitative variables number of employees, turnover, exports and total expenditure for research and innovation the authors build a regression model for each of these variables and release the fitted values. To further protect the tails, i.e. the outlying enterprises, some enterprises are then shrunk towards the centre.

Hereafter we include in the data matrix X the unit vector $\mathbf{1}$, in order to allow for an intercept in the regression. The basic idea is as follows: given a set of variables $X_{K \setminus l}$ and X_l, $l \in K' \subseteq K$, with X_l to be protected, the authors propose to regress X_l on $X_{K \setminus l}$ and to release

$$\tilde{X}_l = \hat{X}_l + a , \tag{1}$$

where $\hat{X}_l$ is the fitted value and $a = sF$ is an adjustment in which s is the predictive standard error and F depends on the rank of X_l. In [11], for a fixed value $p \in (0, 0.5)$, F is taken to decrease linearly from a given value $F_{\max}$ to 0 as the rank of the value of X_l increases from 1 to $[pn]$, to be 0 for values of the rank between $[pn] + 1$ and $n - [pn]$, and to decrease linearly from 0 to $-F_{\max}$ as the rank increases from $n - [pn] + 1$ to n, where $[pn]$ signifies the nearest integer less than pn. Throughout we set $p = 0.25$ and $F_{\max} = 2$. In this way the released values of the first (last) quartile are inflated (deflated) with respect to the corresponding fitted values, with the more extreme values receiving the more extreme inflation or deflation. This tail shrinkage is not applied to values that would already be shrunk if the fitted values were to be released; such a selective tail shrinkage is used in order to reduce error.

The authors perform this regression procedure for each X_l to be protected. Not all variables in the regressions are necessarily protected, therefore we used the notation $l \in K' \subseteq K$. In order to allow for spatial dependence, all of the regressions contain a fixed area effect. For details, see [11].

The method is therefore designed to modify the marginal distributions, redistributing the tail units in the central body of the distribution. This mechanism clearly changes the marginals, but its additional aim is to allow the users to build regression models and to draw almost the same inferential conclusions as they would from the real data. Information about the original marginal distributions and tables can be recovered from the tables published by the NSIs in aggregate form, or can otherwise be supplied to the user.

We follow the same approach as in [11] and apply the same methodology to the Italian sample from the CIS survey, fitting a separate set of regressions plus shrinkage for each economic activity as defined by the two digit NACE classification. We omit enterprises having zero exports and those having innovation at zero cost, because the release of data about these requires special consideration.

We follow [11] and adopt a log-transformation for the numeric variables involved in the regressions because of the skewed nature of the original data.

3.3   Comments on the Protection Method

In this section we discuss the strengths and weaknesses of the protection model discussed in Section 3.2 in light of the application of this technique to the Italian section of the CIS survey.

In particular, we analyse the following issues: flexibility of the procedure; protection achieved; data validity, concerning: means, variances and correlations, regression.

All of the results presented in the sequel pertain to log-transformed data.

Flexibility. First of all, the main feature of the protection model above is flexibility. For example, if we assume that the risk of disclosure is proportional to the distance to the mean of the data, units may be protected according to their risk. Moreover, in formula (1), which specifies the protection model, adjustment terms a can also be constructed such that particular units receive a higher level of protection.

Another positive characteristic of the method is its wide applicability, the core of the method being a set of regression models. Further, such protection strategy is easily implementable in a fully automated fashion.

Protection. By construction, the model is designed to shrink the units towards the mean: in fact, use of the fitted values from the regression models and subsequent modification of the tails both serve to shrink the values towards the mean. This mechanism will in general modify the values of the outlying units.

Figure 1 illustrates the application of principal components analysis to the original and released data. It can be seen that units that are clearly visible in the first graph are effectively moved towards the centre of the data. However, the possibility that some units that fall outside the main body of the released data stay as outliers still remains. In effect, protecting some units may expose others to the risk of disclosure. In cases like this, the best strategy would probably be to re-run the model, using a higher value for p, or otherwise re-run the model on the previously protected data.

Data Validity: Means. Linear regression has the property that means are left unchanged. Moreover, the use of symmetric adjustments a maintains this feature. Note though that when selective shrinkage is adopted, the symmetry just mentioned may disappear, and the mean values might change. However this effect should in general not be dramatic. The computations carried on the real data show a good agreement between the means of the real data and the protected data. Table 1 shows the results for two selected economic activities, NACE rev.1 categories 18 (clothing manufacture) and 24 (chemical products), and for the whole sample of all economic activities.

Table 1. Effect of protection method on the means - selected economic activities

variable	NACE 18		NACE 24		all	
	original	protected	original	protected	original	protected
turnover	10.78	10.75	11.28	11.26	10.08	10.03
exports	8.06	7.99	9.63	9.59	8.38	8.30
# employees	4.57	4.56	5.22	5.22	4.48	4.47
R & I	3.79	3.85	5.60	5.67	3.87	4.94

Data Validity: Variance and Correlation. In general, one side effect of the regression model plus tail shrinkage protection strategy is a certain amount of reduction in the variability. This effect is clearly visible in Figure 2.

From the point of view of an analyst wanting to use the released data to build and estimate regression models, this effect turns into a reduction of the residual variance and results in an apparently increased precision of the estimates. A similar effect is seen for the correlation matrix. Table 2 show some results for food enterprises (NACE code 15) and for the whole sample. In general, the regression acts to strengthen the linear relationships between variables, the only exception being with expenses for research and innovation (R & I); this is probably due to the presence of structural zeroes for those enterprises which, being not engaged in innovation, do not present any expenses for R & I.

Data Validity: Regression. A simple test model for the variable turnover (here denoted by X_2) has been applied to the protected data, divided according to the NACE classification. The design matrix $X_{K\backslash 2}$ contains explanatory variables number of employees and geographical area. For reasons of comparability, the geographical aggregation produced by using principal components analysis

original data, NACE= 18

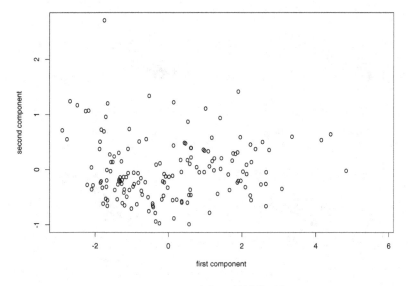

protected data, NACE= 18

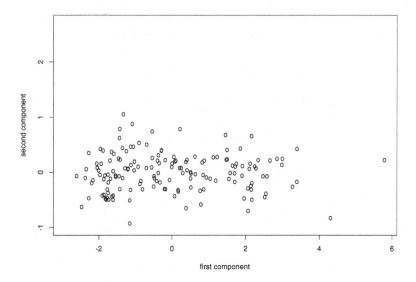

Fig. 1. Effect of the protection method on outlying enterprises. The data have been projected onto their first two principal components. Many outliers have disappeared, although some still remain

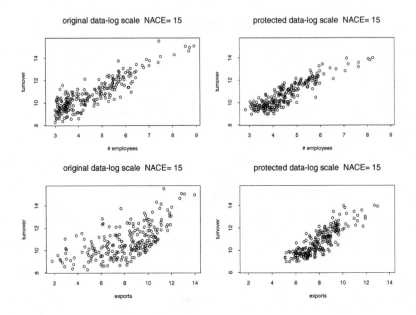

Fig. 2. Effect of the protection method on variability and bivariate relationships

Table 2. Correlation matrices. Correlations for the original data are above the diagonal, correlations for the protected data are below the diagonal

NACE 15			
turnover	-	0.64	0.88	0.45
# employees	0.85	-	0.57	0.30
exports	0.90	0.88	-	0.48
R & I	0.31	0.23	0.34	-
all NACEs			
turnover	-	0.73	0.90	0.46
# employees	0.81	-	0.67	0.37
exports	0.91	0.84	-	0.46
R & I	0.23	0.19	0.23	-

has been used in both models. For enterprises performing the same two digit NACE rev.1 main economic activity, the test model takes the following form:

$$X_2 = X_{K\backslash 2}\beta + \epsilon \qquad \epsilon \sim N(0, \sigma_\epsilon^2 I)$$

for the real data, and

$$\tilde{X}_2 = \tilde{X}_{K\backslash 2}\tilde{\beta} + \eta \qquad \eta \sim N(0, \sigma_\eta^2 I)$$

for the protected data.

The above mentioned drop in variability is again present, as can be seen from Table 3. The table shows estimates of the variance for the following economic activities: 15 (food and beverage), 18 (clothing manufacture), 24 (chemical products) and 28 (metal products).

Table 3. Variance estimates for selected economic activities

NACE	$\widehat{\sigma_\epsilon^2}$	$\widehat{\sigma_\eta^2}$
15	0.67	0.52
18	0.90	0.61
24	0.53	0.42
28	0.49	0.38

Within the previous framework as above, an alternative approach replacing the fitted values by the fitted values themselves plus a random term drawn from the estimated error distribution might provide a solution to the problem of reduced residual variability. Note however that, due to the explained variation, the variability of $\hat{X}_l$ as predicted by a regression model is generally smaller than the variability of X_l. We therefore suggest that a variance inflation factor for the estimators should be released together with the data.

The reduction in variance will of course be present only for the imputed variables. The survey specific variables will be released unchanged.

Having modified the data, one cannot expect the parameter estimates for test regressions based on the released data to be the same as the estimates based on the original data. However, we hope that, on average, the fitted values based on the protected data will be the same as those based on the original data. To see this, consider for simplicity a regression imputation strategy without tail shrinkage. In the notation introduced in Section 3.2, we fit the following models:

$$E[X_l] = X_{K\backslash l}\beta \qquad (2)$$

for each $l \in K'$. We then release

$$\tilde{X} = (\hat{X}_{K'}, X_{K\backslash K'}) ,$$

where $\hat{X}_{K'}$ is a matrix of fitted values obtained from the regressions. Extending to the released matrix $\tilde{X}$ the notation used so far for the original matrix X, with

$\tilde{X}_K$ we define the set of released variables involved in the regression. An analyst building the same model as in (2) for variable l will use:

$$\hat{X}_l = \tilde{X}_{K\backslash l}\alpha + \tilde{\epsilon} \ . \tag{3}$$

According to the standard assumptions of the normal regression model, we have, from model (2), that $E[\hat{X}_l] = X_{K\backslash l}\beta$. From (3) we have that $E(\hat{X}_l) = \tilde{X}_{K\backslash l}\alpha$, whence $\alpha = (\tilde{X}'_{K\backslash l}\tilde{X}_{K\backslash l})^{-1}\tilde{X}'_{K\backslash l}X_{K\backslash l}\beta$. In the absence of tail shrinkage, the expected values of the fitted values for $\tilde{X}_l$ would hence be the same as those for X_l. Both inferences would involve the same regression curve, even though the regression coefficients would differ unless the design matrices coincide: $\tilde{X}_{K\backslash l} = X_{K\backslash l}$.

Empirical evidence shows that in the majority of cases the shrinkage applies symmetric weights to the tail units, the only exceptions resulting from what we termed selective shrinkage; consequently the effect of shrinkage should not be dramatic.

4   Concluding Remarks and Further Research

In this paper we propose a model based framework for disclosure protection methods. We argue that each protection method has its own reference model. According to the number of restrictions imposed, we then classify these methods into nonparametric, semiparametric and fully parametric. Moreover we present the results of an extensive study on application of a protection technique based on regression model [11]. We highlight positive features, limitations, and issues to be further investigated in order to implement such method in the software Argus currently developed as part of the project Computational Aspects of Statistical Confidentiality (CASC).

Throughout the discussion (Section 3.3) we saw that the protection model acts in the sense of strengthening the relationships predicted by the regression models used. Consequently, we must be careful to include in the regression the important predictors, and especially any nonlinear relationship, as the imputation strategy tends to throw away any structure not included in the one imposed by the model itself. A good fit is of course a basic requirement the protection model should meet in order to be sensible and effective; a model exhibiting a poor fit might in fact preserve only a negligible portion of the information carried by the data.

We may use a form of the perturbation function F that differs from the piecewise linear one that we have used. Moreover, in its present form, F depends on the units only through their ranks; using a function of the residuals might link the shrinkage to the (local) fit of the model. This link should be based on optimisation algorithms whose formalisation is a difficult task.

With the present definition of the tail shrinkage function, selection of $F_{\max}$ is crucial and should be tailored to the regression models used and the desired protection level. For example, in order to further protect units that are still at risk after the protection step, we might think of running the method again, using a higher value of F_{max}. Of course, too high a value for F_{max} would shrink the data too much, and vice versa for too small a value of F_{max}.

The normal regression model is sensitive to the presence of outliers. When analysing data like the business microdata, which are characterised, by their nature, by the presence of outliers, use of robust models such as least absolute deviation regression or trimmed regression would be advisable. Indeed, these methods depend less heavily on outlying points, and more importantly, on influential observations.

Evaluation of a variance inflation factor for the estimators is another point deserving further investigation. The major difficulty is the need to devise a factor taking into account also the effect of the shrinkage component.

Provided that the limitations encountered in the application are overcome, we are confident that the method will provide a valuable solution for the dissemination of business microdata.

Acknowledgements

The authors would like to thank Giulio Perani for providing the data, and Giovanni Merola and Giovanni Seri for helpful comments.

This work was partially supported by the European Union project IST-2000-25069 CASC on Computational Aspects of Statistical Confidentiality.

The views expressed are those of the authors only and do not necessarily reflect the policies of the Istituto Nazionale di Statistica or the University of Plymouth.

References

1. Brand, R.: Microdata protection through noise addition. In: "Inference Control in Statistical Databases", LNCS 2316, Springer-Verlag (2002), 97–116.
2. Cox, L.H.: Matrix masking methods for disclosure limitation in microdata. Surv. Method. **20** (1994) 165–169.
3. Cox, L.H.: Towards a Bayesian Perspective on Statistical Disclosure Limitation. Paper presented at ISBA 2000 - The Sixth World Meeting of the International Society for Bayesian Analysis (2000).
4. Dandekar, R., Cohen, M., Kirkendall, N.: Applicability of Latin Hypercube Sampling to create multi variate synthetic micro data. In: ETK-NTTS 2001 Preproceedings of the Conference. European Communities Luxembourg (2001) 839–847.
5. Dandekar, R., Cohen, M., Kirkendall, N.: Sensitive micro data protection using Latin Hypercube Sampling technique. In: "Inference Control in Statistical Databases", LNCS 2316, Springer-Verlag (2002), 117–125.
6. Duncan, G.T. and Mukherjee S.: Optimal disclosure limitation strategy in statistical databases: deterring tracker attacks through additive noise. J. Am. Stat. Ass. **95** (2000) 720–729.
7. Dempster, A.P., Laird, N.M., Rubin, D.B.: Maximum likelihood from incomplete data via the EM algorithm. J. Roy. Stat. Soc. B **40** (1977) 1–38.

8. Domingo-Ferrer, J., Mateo-Sanz, J.M.: Practical data-oriented microaggregation for statistical disclosure control. IEEE Transactions on Knowledge and Data Engineering In Press (2001).
9. Fienberg, S.E., Makov, U., Steele, R.J.: Disclosure limitation using perturbation and related methods for categorical data (with discussion). J. Off. Stat. **14** (1998) 485–502.
10. Franconi, L., Stander, J.: Model based disclosure limitation for business microdata. In: Proceedings of the International Conference on Establishment Surveys-II, June 17-21, 2000 Buffalo, New York (2000) 887–896.
11. Franconi, L., Stander, J.: A model based method for disclosure limitation of business microdata. J. Roy. Stat. Soc. D Statistician **51** (2002) 1–11.
12. Fuller, W.A.: Masking procedures for microdata disclosure limitation. J. Off. Stat. **9** (1993) 383–406.
13. Grim, J., Boček, P., Pudil, P.: Safe dissemination of census results by means of Interactive Probabilistic Models. In: ETK-NTTS 2001 Pre-proceedings of the Conference. European Communities Luxembourg (2001) 849–856.
14. Kennickell, A.B.: Multiple imputation and disclosure protection. In: Proceedings of the Conference on Statistical Data Protection, March, 25-27, 1998 Lisbon (1999) 381–400.
15. Kim, J.: A method for limiting disclosure of microdata based on random noise and transformation. In: Proceedings of the Survey Research Methods Section, American Statistical Association (1986) 370–374.
16. Little, R.J.A.: Statistical analysis of masked data. J. Off. Stat. **9** (1993) 407–426.
17. Little, R.J.A., Rubin, D.B.: Statistical Analysis with Missing Data. John Wiley New York (1987).
18. Raghunathan T., Rubin, D.B.: Bayesian multiple imputation to Preserve Confidentiality in Public-Use Data Sets. In: Proceedings of ISBA 2000 - The Sixth World Meeting of the International Society for Bayesian Analysis. European Communities Luxembourg (2000).
19. Rubin, D.B.: Discussion of "Statistical disclosure limitation". J. Off. Stat. **9** (1993) 461–468.
20. Winkler, W.E., Yancey, W.E., Creecy, R.H.: Disclosure risk assessment in perturbative microdata protection via record linkage. In: "Inference Control in Statistical Databases", LNCS 2316, Springer-Verlag (2002), 135–152.

Microdata Protection through Noise Addition

Ruth Brand

Federal Statistical Office of Germany, 65180 Wiesbaden
ruth.brand@destatis.de

Abstract. Microdata protection by adding noise is being discussed for more than 20 years now. Several algorithms were developed that have different characteristics. The simplest algorithm consists of adding white noise to the data. More sophisticated methods use more or less complex transformations of the data and more complex error-matrices to improve the results. This contribution gives an overview over the different algorithms and discusses their properties in terms of analytical validity and level of protection. Therefore some theoretical considerations are shown and an illustrating empirical example is given.

Keywords: Statistical disclosure control, microdata protection, noise addition.

1 Introduction

Statistical disclosure control is part of the usual preparations necessary for disseminating microdata. Usually direct identifiers and exposed cases are removed, variables that can be used for re-identifications are recoded or dropped and sub-samples are drawn. For some surveys these methods are not sufficient. Therefore other methods, so called perturbation methods, were developed. These methods should assure a high level of protection in surveys that can not be protected sufficiently by the methods mentioned above.

Data perturbation techniques can be divided in two main categories: methods that assure anonymity during the interview (randomized response techniques, for example [3]) and methods that can be interpreted as part of the editing-process of surveys necessary for disseminating files. Literature on the latter methods focuses on techniques like Resampling, Blanking and Imputation, Data-Swapping, Post-Randomization, Microaggregation, Adding noise [6,8]. These techniques differ substantially as to the statistical methods they are based on. This contribution focuses on one of these approaches (namely masking by adding noise). It gives an overview over the algorithms used with respect to their analytical usefulness and compares the protection levels achieved for an empirical example.

For examining analytical usefulness one has to decide whether the descriptive statistics of the original sample should be replicated exactly by the disseminated data set or whether it is sufficient to preserve them in terms of expected values. For scientific use files it seems to be necessary to preserve the possibility to obtain unbiased or at least consistent estimates of central sample statistics, because

J. Domingo-Ferrer (Ed.): Inference Control in Statistical Databases, LNCS 2316, pp. 97–116, 2002.
© Springer-Verlag Berlin Heidelberg 2002

empirical sociological or economic studies usually evaluate causal hypothesis by multivariate statistics or econometric analysis[1].

Central sample statistics needed for applying most multivariate methods are the sample means and the sample covariance of the masked data. It should be possible to derive consistent estimators for the true means and covariance in terms of the masked variables for assuring that standard-techniques, especially OLS-regression-estimates, can be applied to the anonymized data. Nevertheless the calculated values should be close to those of the original data to obtain similar results [15,16]. Furthermore it is desirable to preserve at least the univariate distributions due to their usefulness for a general description of the data and the deduction of analytical models.

Despite this statistical disclosure control methods have to assure confidentiality of the individual responses. A detailed analysis of this aspect would have to take into account two components: First the direct reduction of the re-identification risk or the risk of an inferential disclosure. Second the information loss connected to perturbation of those variables that contain the information an intruder is interested in. These aspects are not discussed in detail. The reason for this is that the properties of the different algorithms have an important impact on the results - an aspect that can not be considered in this contribution. Therefore the level of protection is only highlighted by a short summary of the literature and an empirical example.

This contribution gives an overview over the main methods of masking by adding noise. At first simple addition of random noise is described. Second the algorithm conducted by Kim [10] is examined and third a different approach conducted by Sullivan [21,7] is discussed. Fourth an empirical example is given that highlights the properties of the algorithms and allows a short comparisons of the protection level.

2 Description of Algorithms

2.1 Adding Noise

Masking by adding noise was first tested extensively by Spruill [20]. (see also [18,10,24,25]). The basic assumption is, that the observed values are realizations of a continuous aleatory variable: $x_j \sim (\mu_j, \sigma_j^2)$, $j = 1, \ldots, p$. Adding noise means that the vector x_j is replaced by a vector z_j:

$$z_j = x_j + \epsilon_j \, , \tag{1}$$

where $\epsilon_j \sim N(0, \sigma_{\epsilon_j}^2)$, $j = 1, \ldots, n$, denotes normally distributed errors with $Cov(\epsilon_t, \epsilon_l) = 0$ for all $t \neq l$ (white noise) generated independently from the

---

[1] Explicit estimates of population values are seldom focused in empirical analyzes published in scientific journals. Furthermore dissemination of sampling weights increases re-identification risks due to their deduction from the sample design [27,4]. However it is assumed that the disseminated dataset contains no weights or similar information about the sample design.

original variables, which implies $Cov(x_t, \epsilon_l) = 0$ for all t, l. Using matrices this can be written as:

$$Z = X + \epsilon \ ,$$

with $X \sim (\mu, \Sigma)$, $\epsilon \sim N(0, \Sigma_\epsilon)$, and Z as matrix of perturbated values:

$$Z \sim (\mu, \Sigma_Z) \quad , \quad \text{with} \quad \Sigma_Z = \Sigma + \Sigma_\epsilon \ . \tag{2}$$

Furthermore the general assumption in literature is that variances of the ϵ_j are proportional to those of the original variables[2] [20,24,25]. This leads to:

$$\Sigma_\epsilon = \alpha \ diag(\sigma_1^2, \sigma_2^2, \ldots, \sigma_p^2) \ , \quad \alpha > 0 \ , \tag{3}$$

where $diag(\sigma_1^2, \sigma_2^2, \ldots, \sigma_p^2)$ denotes a diagonal matrix which has the variances on the main diagonal and all off diagonal elements zero and α a positive constant used for varying the "amount of noise". Assuming that the error-variance is proportional to the variance of the corresponding variable assures that the relative error $(\sigma_x^2/\sigma_\epsilon^2)$ is identical for all variables whereas the independence assumption results only from the fact that the variables are described isolated in literature [25]. Under these conditions we obtain for the masked data:

$$\Sigma_Z = \Sigma + \alpha \ diag(\sigma_1^2, \sigma_2^2, \ldots, \sigma_p^2) \quad . \tag{4}$$

This method leads to useful results if relatively small errors are used [20]. This can be explained with the structure of errors. Obviously the sample means of the masked data are unbiased estimators for the expectancy values of the original variables due to the error-structure chosen: $E(Z) = E(X) + E(\epsilon) = \mu$. Nevertheless this is not true for the covariance matrix: Although the covariances are identical $Cov(z_j, z_l) = Cov(x_j, x_l)$, for all $j \neq l$, the variances of z_j and x_j differ:

$$V(z_j) = V(x_j) + \alpha V(x_j) = (1 + \alpha)V(x_j) \ . \tag{5}$$

This implies that the sample variances of the masked data are asymptotically biased estimators for the variances of the original data. However the sample covariances are consistent estimators for the covariances of the original data.

For the correlation between z_l and z_j this leads to:

$$\rho_{z_l, z_j} = \frac{Cov(z_l, z_j)}{\sqrt{V(z_l)V(z_j)}} = \frac{1}{1+\alpha} \frac{Cov(x_l, x_j)}{\sqrt{V(x_l)V(x_j)}} = \frac{1}{1+\alpha} \rho_{x_l, x_j} , \quad \forall \ j \neq l \ .$$

Obviously sample correlations of the masked variables are biased estimators for ρ_{x_l, x_j}. They are inconsistent too, because this bias will not vanish with an increasing sample size. Nevertheless if α is known estimates can be adjusted.

---

[2] In practice the covariance matrix of X (Σ) is not known and must be estimated by its sample counterpart before the errors are generated. Due to the fact that the sample means and covariances are consistent estimators for the corresponding central moments the use of sequential expectations allows a generalization of the results [10].

Due to these findings it is suggestive to generate an error matrix ϵ^* under the restriction $\Sigma_{\epsilon^*} = \alpha\Sigma$ (correlated noise), which implies:

$$\Sigma_Z = \Sigma + \alpha\Sigma = (1+\alpha)\Sigma \ . \tag{6}$$

In this approach all elements of the covariance matrix of the perturbed data differ from those of the original data by factor: $1 + \alpha$. The leads to:

$$\rho_{z_j,z_l} = \frac{1+\alpha}{1+\alpha} \ \frac{Cov(x_j,x_l)}{\sqrt{V(x_j)V(x_l)}} = \rho_{x_j,x_l}$$

for all $j = 1,\ldots,p$ and $l = 1,\ldots,p$, correlations of the original data can be estimated asymptotically unbiased by the sample correlations of the masked data.

In a regression framework adding correlated noise leads to unbiased estimates for the coefficients. Nevertheless the residual variance calculated from an regression estimate based on the masked data is an asymptotically biased estimate of its unmasked counterpart. Due to the fact that the residual variance is a central element of the coefficient of determination and several tests usually used for evaluating goodness of fit this is a drawback [1]. However the researcher can adjust for this bias as long as α is known.

Kim [11] shows that expected values and covariances of subpopulations can be estimated consistent as long as α is known. Furthermore it is possible to obtain consistent estimates of regression parameters when only some of the variables used are masked (partial masks, see [11]). These findings illustrate a general advantage of this method: Consistent estimators for several important statistics can be achieved as long as the value of parameter α is known. This means identical results can be achieved on the average, numerical equivalence of every estimate is not implied.

A general disadvantage that can not be adjusted by 'external' information about the masking procedure is that the distributions of the masked variables can not be determined as long as the variables x are not normally distributed. The reason for this is that the distribution of the sum of a normally distributed variable (the error-term added) and a not normally distributed variable x_j is not known in general.

Summarizing simple masks by adding correlated noise seems to be preferable because of its pleasant properties for analysis. Nevertheless it is usually not used due to its very low level of protection (see e.g. [12,24,25]). Therefore it is primarily a reference framework for studying general problems of adding noise to continuous variables.

2.2 Masking by Adding Noise and Linear Transformations

On the basis of the results for the simple mask Kim [10] proposes a method that assures by additional transformations that the sample covariance matrix of the masked variables is an unbiased estimator for the covariance matrix of the original variables (X)[3].

This method bases on a simple additional mask: $z_j = x_j + \epsilon_j$, $j = 1, \ldots, p$, with error-covariances proportional to those of the original variables (eq. (6)). Kim [10] proposes to choose either a normal distribution or the distribution of the original variables as error-distribution[4].

In a second step every overlayed variable z_j is transformed:

$$g_j = cz_j + \iota d_j , \qquad j = 1, \ldots, p ,$$

where g_j denotes the overlayed and transformed variable (masked variable) and ι is a vector of ones. The scalar parameter of the transformation c is constant for all variables, whereas the parameter d_j differs between the variables. Formulating the procedure in matrices leads to:

$$Z = X + \epsilon$$
$$G = cZ + D$$
$$= c(X + \epsilon) + D ,$$

where X denotes the matrix of original variables, $X \sim (\mu, \Sigma)$, ϵ a matrix of errors for masking (overlays), $\epsilon \sim (0, \alpha\Sigma)$, G the matrix of overlayed and transformed variables, $G \sim (\mu, \Sigma)$, and D denotes a matrix that has the typical element $d_{tj} = d_j$.

The parameters c and d_j are determined under the restrictions: $E(g_j) = E(x_j)$ and $V(g_j) = V(x_j)$ for all j, $j = 1, \ldots, p$ [10]. The first restriction implies:

$$d_j = (1 - c)E(x_j) ,$$

which leads to two possible transformations:

$$g_{j,1} = cz_j + (1 - c)\bar{x}_j \qquad (7)$$
$$g_{j,2} = cz_j + (1 - c)\bar{z}_j . \qquad (8)$$

Equation (8), which is chosen by [10], leads to:

$$c = \sqrt{\frac{n - 1 - \alpha}{(n - 1)(1 + \alpha)}} , \qquad (9)$$

using the restriction $\sigma_X = \sigma_G$ which implies $\Sigma_X = \Sigma_G$. For large samples Kim [10] proposes to use the limit:

$$\lim_{n \to \infty} c = \frac{1}{\sqrt{1 + \alpha}} .$$

---

[3] The same scheme is proposed by Tendick and Matloff [26] with respect to query processing in data base management systems.

[4] For practical purposes normal distributions are usually used in literature, see [5,12,13,14,17,29]. Roque [19] proposes mixtures of multivariate normal noise.

Using equation (7) for determining the parameter c leads to [1]:

$$c = \sqrt{\frac{n-1}{n(1+\alpha)-1}} \ .$$

which is asymptotically equivalent to (9). Furthermore it can be shown easily that the results do not differ significantly in small samples.

Due to $\alpha > 0$ parameter c takes values between zero and one. Therefore the masked variables g_j can be interpreted as weighted averages of the overlayed values d_j and the corresponding means ($\bar{x}_j$ or $\bar{d}_j$). For "small" values of c the masked data points are compact around the mean while the correlation structure is not affected on the average. For a given sample size c depends strongly on the relative amount of noise (α), with increasing α the value of c decreases.

This algorithm preserves expectancy values and covariances due to the restriction used for determining transformation parameter c, this means: $E(G) = E(X)$ and $E(\hat{\Sigma}_G) = \Sigma_X$. On this basis Kim [10,11,24] derives several results for linear regression analysis. Obviously coefficients and their variances can be estimated consistently as well as the variance or the regression-errors, because all covariance matrices of a model formulated in the original variables can be estimated consistently by their counterparts in the masked data [10,11]. The same can be shown for partial masks [11]. These results are confirmed empirically by several examples in [10,11,12,13]. They show that the algorithm often leads to quite good results in terms of analytical validity.

Furthermore Kim [11] considers the analysis of regression estimates in sub-populations. In [11] it is shown that sample means and sample covariances are asymptotically biased estimates of the interesting moments. The magnitude of the bias depends on parameter c. Therefore estimates can be adjusted as long as c is known.

Another property often relevant for empirical analysis is the univariate distribution. This is not sustained in general by this algorithm. This is reasoned (again) by the fact that the sum of two random variables does not follow the distribution of one of those in general. Only if the original variables are normally distributed the sum is normally distributed, too.

Summarizing the algorithm proposed by Kim [10] seems to be suitable for continuous variables: It preserves central statistics like the correlation matrix and expected values or allows deviation of adjusted estimates. Furthermore partial masks are possible and sub-populations can be analyzed. From this point of view it is a convenient method for anonymizing microdata. Nevertheless the univariate distributions are not sustained. Another difficulty is that it is not applicable for discrete variables due to the structure of the transformations.

2.3   Masking by Adding Noise and Nonlinear Transformation

Sullivan [21] proposes an algorithm that combines a simple mask with nonlinear transformations. Furthermore it can be applied to discrete variables. The univariate distributions and the first and second empirical moments is explicitly considered.

This method consists of several steps [7,22]:

1. Calculating the empirical distribution function for every variable,
2. Smoothing the empirical distribution function,
3. Converting the *smoothed* empirical distribution function into a uniform random variable and this into a standard normal random variable.
4. Adding noise to the standard normal variable,
5. Back-transforming to values of the distribution function,
6. Back-transforming to the original scale.

During the application of the algorithm a distance criterion is used. It assures that the distance between the masked variables obtained in step four and its "standard normal" counterpart is not one of the two smallest distances. The properties of this more complicated algorithm are determined by the application of all steps. Therefore they will be described in this section in more detail.

Steps 1 and 2 contain different transformations for continuous and discrete variables. For a continuous variable X the empirical distribution is calculated in step 1. In step 2 this distribution is transformed to the so called smoothed empirical distribution [21]. Therefore the midpoints between the values X_i will be used as domain limits: $x_i = (X_i + X_{i+1})/2$ for all $i = 0, \ldots, m$, with $X_0 = 2X_1 - X_2$, $X_{m+1} = 2X_m - X_{m-1}$. Within these classes the smoothed empirical distribution is calculated:

$$\bar{F}_x(z) = \hat{F}(x_{i-1}) + \frac{\hat{F}(x_i) - \hat{F}(x_{i-1})}{x_i - x_{i-1}}(z - x_{i-1}) \qquad \text{for} \quad z \in (x_{i-1}, x_i] \ , \quad (10)$$

where x_i denotes the domain limits and $\hat{F}(x_i)$ the values of the empirical distribution function at the limits. Equation (10) is calculated for every value of X: $p_i = \bar{F}_x(X_i)$. These are mapped into standard normal values by applying the quantile function:

$$Z_i = \Phi^{-1}(p_i) \ .$$

These transformations are a standardization for normally distributed variables due to: $\hat{F}(X_{ij}) \simeq \Phi(X_{ij})$. The correlations of the transformed variables are nearly identical to those of the original variables as long as the variables are jointly normal.

If the observed variables are not normally distributed the correlations of the transformed variables differ substantially. The magnitude depends on the differences between the empirical distribution of standardized values and the standard normal distribution.

For transforming a discrete variable with k possible outcomes it is first split in $k-1$ Bernoulli variables [23]. Second the conditional covariance matrix of the dummy-variables, given the continuous variables ($m_{dd.cc}$) is calculated:

$$m_{dd.cc} = m_{dd} - m'_{cd}m_{cc}^{-1}m_{cd} \ , \tag{11}$$

where m_{cc} denotes the covariance matrix of the continuous variables, m_{dd} the covariance matrix of the binary variables and m_{cd} the matrix of covariances between continuous and binary variables.

Third a matrix of standard normal random numbers F_{dc} is generated, with column vector f_{dct}:

$$f_{dct} = m'_{cd} m_{cc}^{-1} L_{cc}^2 Z_{ct} + m_{dd.cc}^{1/2} e_{d.ct} , \tag{12}$$

where Z_{ct} denotes the vector of transformed continuous variables stemming from observation t, $L_{cc}^2 = diag(m_{cc})$ a matrix that has the sample variances on the diagonal and all off-diagonal elements zero and $e_{d.ct}$ a vector of standard normal random numbers. It is easily seen that the f_{dct} are approximately normal with expected value 0 and covariance matrix m_{dd}. Furthermore the pair (Z_{ct}, f_{dct}) is approximately normal and has nearly the same correlation matrix as the original data.

Although the f_{dct} have nearly the same correlations as the original Bernoulli variables they do not depend on their true values. To link the original values of the Bernoulli variables with f_{dct} further transformations are needed. Therefore the values of the distribution function are determined for all $j = 1, \ldots, p_{d,}$:

$$g_{dct,j} = \Phi(f_{dct,j}) .$$

The $g_{dct,j}$ are realizations of uniform random variables. Nevertheless they do not depend on the Bernoulli variables either. Hence the random variable $h_{dct,j}$ is generated that depends on $g_{dct,j}$ and the original variable $(x_{dt,j})$:

$$h_{dt,j} = \begin{cases} g_{dtc,j}(1 - p_{oj}) & \text{if } x_{dt,j} = 0 , \\ 1 - p_{oj}(1 - g_{dtc,j}) & \text{if } x_{dt,j} = 1 \end{cases} \quad j = 1, \ldots, p_d , \tag{13}$$

where p_{oj} denotes the mean of the j-th Bernoulli-variable. Next normalized ranks $R_{d1,j}, \ldots, R_{dn,j}$ are assigned starting with the smallest value of $h_{dt,j}$:

$$\tilde{R}_{dt,j} = \frac{R_{dt,j} - 0.5}{n} .$$

These are transformed in standard normal variables using the quantile function:

$$Z_{dt,j} = \Phi^{-1}(\tilde{R}_{dt,j}) .$$

Combining these with the transformed continuous variables leads to the vector of standard normal variables for every observation: $Z_t = (Z'_{ct}, Z'_{dt})$.

In the fourth step of the algorithm the transformed variables Z_t were masked by adding noise. This mask is in principle identical to the simple mask described in section 2.2. Let Z denote the matrix of transformed standard normal variables with row vectors Z_t and U^* the matrix of errors $U^* \sim N(0, \alpha M_{ZZ})$, the matrix of masked variables Z^* is defined as:

$$Z^* = Z + U^* = Z + \sqrt{\alpha} \, U T_{ZZ} , \quad \text{for } \alpha > 0 ,$$

where $U \sim N(0, I_{p \times p})$ and T_{ZZ} denotes a decomposition of the correlation matrix of the transformed data (P_{ZZ}): $T'_{ZZ} T_{ZZ} = P_{ZZ}$. Since the elements of Z and U^* are normally distributed, the masked values Z_t^* are normal: $Z^* \sim N(0, M_{Z^* Z^*})$, with $M_{Z^* Z^*} = M_{ZZ} + \alpha M_{ZZ} = (1 + \alpha) M_{ZZ}$.

Subsequent for any masked observation a vector of Mahalonobis distances between the masked and all original observations is calculated (m_i) that has the typical element:

$$m_{it} = d_{it} \Sigma_d^{-1} d'_{it} \quad , \tag{14}$$

where d_{it} denotes the vector of differences between observations i and t: $d_{it} = z_i - z_t^*$ that has the covariance matrix: $\Sigma_d = \alpha M_{ZZ}$ [21]. The criterion chosen for sufficiency is that m_{ii} is not one of the two smallest distances in m_i. Otherwise the mask will be repeated until the distance criterion if fulfilled ([21], p.70).

In steps 4 and 5 of the algorithm the masked values are transformed back to the original scale. For each variable Z_j^*, $j = 1, \dots, p$ a vector of normalized ranks D_j^* is calculated. For this purpose a vector of ranks R_j^* is calculated with elements of Z_j^* in ascending order and divided by the number of observations n.

This 'empirical distribution' is modified for back transformation, because its values depend solely on the sample size. Hence the errors are standardized:

$$u_{tj}^+ = \frac{u_{tj}^*}{\sqrt{\frac{1}{n-1} \sum_{t=1}^n u_{tj}^{* \, 2}}} \quad ,$$

mapped into the domain $(0;1)$ and added to the ranks:

$$D_{tj}^* = \frac{R_{tj}^* - \phi(u_{tj}^+)}{n} = \frac{R_{tj}^*}{n} - \eta_{tj} \ , \quad t = 1, \dots, n \ , \tag{15}$$

where $\phi(\bullet)$ denotes the function that maps the u_{tj}^+ to values between zero and one, which implies $0 \le \eta_{tj} \le \frac{1}{n}$. Therefore D_{tj}^* is restricted to:

$$\frac{R_{tj}^*}{n} \ge D_{tj}^* \ge \frac{R_{tj}^* - 1}{n} \ . \tag{16}$$

A suggestive choice for $\phi(\bullet)$ is the standard normal distribution, because u_{tj}^+ is normally distributed which leads to a uniform distributed η_{tj} [21].

The final back transformation differs depending on whether a variable is continuous or discrete. For continuous variables the inverse of the smoothed empirical distribution $\bar{F}_X$ is used:

$$X_{ctj}^* = x_{i-1,j} + \frac{D_{tj}^* - \hat{F}(x_{i-1,j})}{\hat{F}(x_{i,j}) - \hat{F}(x_{i-1,j})} (x_{i,j} - x_{i-1,j}), \text{ for } D_{tj}^* \in (\hat{F}(x_{i-1,j}), \hat{F}(x_{i,j})]. \tag{17}$$

For transforming the binary variables equation (13) is inverted:

$$X_{dtj}^* = \begin{cases} 0 & , \text{ if } D_{tj}^* \in (0, 1 - p_{oj}) , \\ 1 & , \text{ if } D_{tj}^* \in [1 - p_{oj}, 1) , \end{cases} \quad \text{for} \quad t = 1, \dots, n, \quad j = 1, \dots, p_d \ . \tag{18}$$

This back transformations assure that the univariate distributions will be sustained approximately. Therefore sample means and variances are similar to those of the original data. Nevertheless the correlations differ due to numerical differences and/or not jointly normally distributed original variables. Furthermore the correlations between X and X^* differ between the variables, this means that the "relative amount of noise" differs. To adjust for these drawbacks Sullivan [21] (pp. 76) proposes two iterations. First the cross correlations between the variables and their masked counterparts are adjusted to a robust average. Second differences between the elements of the correlation matrices are minimized.

Due to $E(X_j) = E(X_j^*)$ adjusting cross correlations is based on:

$$X_{tj}^* = \rho_{X_j X_j^*} + \eta_{tj} \ ,$$

where $\rho_{X_j X_j^*}$ denotes the correlation between X_j and X_j^* and η_j an error term independent of X_j, $\eta_j \sim (0, \sigma_{\eta_j}^2)$, $Cov(X_j \eta_j) = 0$ with

$$\sigma_{\eta_j}^2 = \sigma_{X_j}^2 (1 - \rho_{X_j X_j^*}^2) \ . \tag{19}$$

This leads to:

$$\rho_{X_j X_j^*}^2 = 1 - \frac{\sigma_{\eta_j}^2}{\sigma_{X_j}^2} \ .$$

It is reasonable to assume that the correlation between the original variables and their masked counterparts is positive, this means $\sigma_{\eta_j}^2 < \sigma_{X_j}^2$. Therefore the correlation increases with declining $\sigma_{\eta_j}^2$. Hence a modification of the error terms can be used for adjusting the correlations respective $\rho_{X_j X_j^*}^2$.

The target correlation chosen is a robust average of the correlations $\bar{\rho}$:

$$\bar{\rho} = \frac{\sum_{j=1}^{p} r_{X_j X_j^*} - (max_{1 \leq j \leq p} \ r_{X_j X_j^*} + min_{1 \leq j \leq p} \ r_{X_j X_j^*})}{p - 2} \ ,$$

where $r_{X_j X_j^*}$ denotes the sample correlation between X_j and its masked counterpart X_j^*.

For determining the amount of variation of variances a simple linear approximation is used. A transformation matrix B_{aa} is defined with typical element:

$$b_{aaij} = \begin{cases} \frac{1 - \bar{\rho}}{1 - 0.5(\bar{\rho} + r_{X_j X_j^*})} & \text{if } i = j \text{ and } i, j = 1, \ldots, p \ , \\ 0 & \text{else} \ . \end{cases}$$

and new standard normal masked values are calculated by:

$$Z_t^* = Z_t + u_t^* B_{aa} \ . \tag{20}$$

These are transformed back to the original scale. This adjustment is done iteratively until the observed cross correlations differ from the desired cross correlations less than a specified amount or until a prespecified number of iterations is exceeded.

As mentioned above correlations of the masked and the original data differ usually. Due to this Sullivan [21], pp. 78 proposes a second iterative adjustment in order to make the off diagonal elements of the correlation matrices nearly identical. The basic idea is again to use a linear transformation for adjusting the error terms. Sullivan ([21], p. 78) proposes to modify them subsequently starting with the variable for which $\sum_k (\rho_{X_j X_k} - \rho_{X_j^* X_k^*})^2$ is largest.

For modifying the errors a linear combination H_1^* of the transformed original variable chosen (Z_1) and the masked values of the other variables (Z_j^*) is created:

$$H_1^* = b_0 Z_1 + \sum_{j=1}^p b_j Z_j^* = Z_1^{+'} b \tag{21}$$

with $b_0 = 1 - b_1$. The system of equations defining the desired properties is $r(G_1^*, X_1) = \kappa$ and $r(G_1^*, X_1^*) = r_{X_1, X_1}$, where G_1^* denotes the back transformed variable corresponding to H_1^* and κ the arithmetic mean of the cross correlations between X_1 and the masked variables $X_2^*, \ldots, X_p^*$: $\kappa = \frac{1}{p-1} \sum_{j=2}^p r_{X_1 X_j^*}$. The coefficients are calculated by solving:

$$\Sigma_{Z^+} b = \rho^+ \ ,$$

where Σ_{Z^+} denotes the correlation matrix of $Z^+ = (Z_1, Z_2^*, \ldots, Z_p^*)$ and $\rho^{+'} = (\kappa, r_{X_1, X_2}, \ldots, r_{X_1, X_p})$. Afterwards the values of H_1^* can be calculated by (21) and transformed back to the original scale by steps 4 and 5 of the algorithm. This approximation can be repeated iteratively until a convergence criterion for the correlations is achieved or a prespecified number of iterations is exceeded.

The algorithm described above allows masking of binary and continuous variables. For discrete variables with more than two categories a final back transformation has to be applied. Let Z_t^* be a vector defined as

$$Z_{ti}^* = \begin{cases} X_{dt1}^* & \text{for } i = 1 \\ (1 - \sum_{i=1}^{i-1} Z_{ti}^*) X_{dti}^* & \text{for } i = 2, \ldots, k-1 \end{cases} \ . \tag{22}$$

Then the elements of the masked variable X_d^* are defined as $X_{dt}^* = i$, if $Z_{ti}^* = 1$.

Summarizing the algorithm proposed by Sullivan [7,21] combines transformations with adding noise. The transformations chosen assure that the univariate distributions are sustained approximately. They are not linear and quite complex in comparison to the ones used by Kim [10,11]. Additionally iterative procedures are used for correcting differences in correlations induced by transformations and mask. Due to these corrections it can not be assured that all variables have the same level of protection.

In order to compare the algorithm with other masking algorithms it has to be analyzed whether the masked variables allow an unbiased estimate of the first two moments of the unmasked variables as long as only binary discrete variables are masked. In [1] it is shown that the sample means of the masked data are unbiased estimates for the expected values of the original data. Furthermore sample variances of masked binary variables are unbiased estimates for the variances of the original binary variables.

Nevertheless the variances of continuous variables increase [1] due to:

$$V(X_{cj}^*|x_{0,j}, x_{1,j}, \ldots, x_{n,j}) = V(X_{cj}) + \sum_{i=1}^{n} \frac{(x_{i,j} - x_{i-1,j})^2}{12n} , \qquad (23)$$

where $x_{i,j}$ denotes the domain limits (see eq. (10)). The increase is higher with rising sample size and larger differences between the ordered observed values in the original data. This result shows that the sample variance calculated by the masked variables $\hat{V}(X_{cj}^*) = \frac{1}{n}\sum_{i=1}^{n}(X_{c,ij}^* - \bar{X}_{cj}^*)^2$ is a biased estimator for the sample variance of the original data. The variation in the original data is overestimated due to the additional variation within the 'classes' used for transformation. Nevertheless the sample variance of the masked data is a consistent estimator for the variance of the original data, because the limit of the term added tends to zero for $n \to \infty$.

Due to the complex structure of the algorithm a more detailed analysis of the properties is not useful. The main reason for this is that the distribution of the errors on the original scale can not be determined explicitly. Corresponding only general inferences for regression estimates are possible on the basis of an errors-in-variables-model [7].

Summarizing the algorithm proposed by Sullivan [7,21,22], can be characterized as a complex method that combines adding noise with nonlinear transformations. Univariate distributions of the variables are sustained approximately. Nevertheless variances of continuous variables increase in small samples due to the structure of the transformations. The algorithm assures that the correlation structure of the original variables is preserved due to iterative adjustment procedures.

3   Empirical Results

In the previous sections three masking algorithms are described and their analytical properties are discussed on the basis of statistical inference. In this section empirical results found in literature are summarized and a numerical example for the masking algorithms described above is discussed. Therefore a subsample of simulated data developed for testing statistical disclosure methods for tabular data is used. These data are characterized by extremely skewed distributions. Furthermore negative values occur and high proportions of the observations have zero values. This characteristics are typical for many business data like data on staff, costs, business taxes and volume of sales.

The subsample drawn contains 1000 'observations' and 7 continuous variables. All variables were divided by factor 100 000 in order to reduce the differences between the absolute values. The descriptive statistics of the prepared dataset can be found in Tables 1 and 2 in the annex. This dataset was masked by four algorithms: white noise, noise with correlation structure identical to sample correlations, the method proposed by Kim [10] and the method proposed by Sullivan [21]. For all masks the value 0.5 is chosen for the 'relative amount of noise' ($\alpha = 0.5$). For the algorithm proposed by Sullivan the target differences for the iterative adjusting procedures are set to 0.01; the maximum number of iterations for adjusting cross correlations between X and X^* (eq. (20)) is set to 50 and the iterations for equalizing the covariances are restricted to a maximum of 1000 for a single variable; after reaching this limit the procedure is aborted.

The results of the masks are shown in Tables 3 - 9 in the annex. Adding white noise has no effects on means and variances; correlations differ clearly as expected. A simple mask with correlated errors inflates variances by factor $1 + \alpha$ while the correlations should not be affected. This is clearly not true for $VAR414$ due to its extreme distribution. The differences in Table 6 occur, because the realizations differ from the ideal orthogonality conditions stated in the assumptions. Due to the high proportion of zeros combined with an extremely skewed distribution this numerical effect leads to serious differences in the correlations for this variable.

For the method proposed by Kim (section 2.2) theoretical findings are confirmed, too. Please notice that the correlations connected with $VAR414$ are again seriously distorted due to realizations of errors. Despite this it has to be stated that all methods discussed (white noise, correlated noise, Kims algorithm) lead to a significant distortion of the distributions including their characteristics like maxima, minim and number of zero values. These findings confirm the results described in literature, e.g. [13,17,20,24], although for Kim's algorithm some of the numerical differences between the correlations are larger than reported in [13]. This is probably reasoned by the extreme distributions of our data in comparison to the income data used in [13].

The algorithm proposed by Sullivan leaves the means nearly unaffected and leads to variances that are larger than those of the original variables. Furthermore minima, maxima and the proportion of values near zero are not too far from the values in the original data due to the transformations chosen (section 2.3). Adjustment of cross correlations terminated at its limit of 50 iterations and the iterations used for equalizing the correlation matrices aborted at their limit for variable $VAR413$ after being successfully used for four other variables. Therefore correlations differ by a maximum difference of 0.031 which is related to the correlation between $VAR413$ and $VAR4$. These results are better than expected. Although the distributions of the original variables are extremely skewed and therefore the transformed standard normal variables have a correlation structure that is different from the original data the adjustment procedures work quite well.

Some examples for application of the algorithm proposed by Sullivan can be found in the literature [1,2,7,21,23] that highlight the properties. These examples confirm that sample means and variances are approximately sustained for different distributions of the original variables. Differences in the correlations occur if the number of iterations in the procedures adjusting the correlations is limited to a few or if the distributions of the original variables are extremely skewed. Nevertheless estimates in subpopulations are biased due to the structure of transformations and an explicit adjustment formula is not known [21,23]. Furthermore for Sullivan's algorithm it can be shown that regression analysis with limited dependent variables can be misleading, because the multivariate distribution is not sustained [1]. The same is true for all methods discussed above if likelihood estimations require an explicit determination of the error distribution.

Summarizing all methods that use correlated noise sustain the correlations approximately although numerical differences occur. In the example given these differences are larger than usually stated in literature due to the extreme distributions in the original data. The algorithm proposed by Sullivan [21] allows adjusting for these differences if the iterative adjustment procedures connected with the mask work[5]. The univariate distributions are solely preserved by the algorithm proposed by Sullivan. Furthermore only this algorithm allows masking of discrete variables. Therefore Kim's algorithm is often combined with additional data swaps [13,17]. Nevertheless it seems to be preferable if analysis of subgroups is required. Hence the methods described can not be ranked. Whether one or the other is appropriate depends strongly on the purposes of potential analyzes.

4   Level of Protection

The level of protection connected to masking methods is widely analyzed in the literature (see e.g. [29,5,17,29]). Nevertheless a standard method that is clearly preferable is not available. Therefore the level of protection will be discussed in a short review of literature and an empirical comparison on the basis of the example given in the foregoing section.

The level of protection achieved by adding correlated noise was first analyzed by Spruill [20] who used an unweighted distance criterion. Spruill comes to the result that 'if there are few common variables, then any releasing strategy that introduces only a small amount of error to the data can be used. If there are more than a few (4 - 6) variables then care must be taken'[20] p. 605. The algorithm proposed by Kim [10] is extensively tested in [17,13,29] using a matching algorithm conducted by Winkler [28]. These studies show that this masking algorithm is not sufficient. Due to this result additional data swaps are proposed [17]. This unsatisfactory result is confirmed by Domingo-Ferrer-Mateo-Sanz [5] who used different methods for assessing the disclosure risk. Tendick [25] shows for normally distributed variables that forecasts based on a linear model formulated in the masked variables may be more precise than forecasts calculated by adding white noise.

The algorithm conducted by Sullivan has not been systematically evaluated until now. The studies available use simple distance criteria like the one used during its application. Calculating the Mahalanobis-distance can be interpreted as a first step to assessing the matching probability for normally distributed variables [21]. If not normally distributed variables are used like in the current example it is a relatively simple procedure for giving a first insight into the

---

[5] The procedures fail, if the initial mask does not fulfill the restrictions described in section 2.3.

problem. Nevertheless re-identification probabilities can not be determined. In the literature available [21,7,1], [22] it is shown that the algorithm masks most observations properly if the described criterion is used.

Although this can not be seen as a sufficient condition for achieving a high protection level [1], the criterion can be used for a short comparison of disclosure risk reduction achieved by the algorithms. Table 10 in the annex shows the results for examining the distances between the original data and their masked counterparts. It is easily seen that correlated noise has a very low level of protection as found in literature. Furthermore the algorithm conducted by Kim leaves a high number of cases "unprotected" in the sense that none or only a few masked observations have smaller distances to the observation looked at than its own mask. Surprisingly adding white noise leads to more 'protected' cases than this technique. The algorithm proposed by Sullivan has the lowest number of "unprotected" cases due to the enclosed criterion for sufficiency. Nevertheless the masked data of the example are not sufficiently masked (measured by the internal criterion), because 62 of the observations do not fulfill the internal criterion formulated for standard normal variables.

These results show that some techniques used for masking by adding noise do not reduce disclosure risks sufficiently. But clear differences between the algorithms can be stated. Hence a systematic evaluation of promising techniques seems to be necessary.

5   Summary

Masking by adding random noise has been discussed for nearly twenty years now. The basic idea is to add independently generated noise to the data or a transformation of them[6].

This contribution gives an overview over these techniques and compares them in terms of analytical usefulness and protection level. Several methods of masking by adding noise that are proposed in literature (see e.g. [10,12,21], [7]) are described. It can be shown that they have very interesting properties with respect to the analytical usefulness of the masked data while the level of protection connected to most techniques seems not to be sufficient. Nevertheless for some techniques a systematic evaluation of the protection level has not yet been achieved. Furthermore the techniques that can not assure a sufficient level of protection may be appropriate if they are combined with other methods.

---

[6] An alternative modeling by multiplying random noise [9,20] can be seen as adding noise to the logarithms of the data. For a generalized discussion see [14,15].

References

1. Brand, R. (2000): Anonymität von Betriebsdaten, Beiträge zur Arbeitsmarkt- und Berufsforschung Beitrag 237, Institut für Arbeitsmarkt- und Berufsforschung, Nürnberg.
2. Brand, R., Bender, S. and S. Kohaut (1999): Possibilities for the Creation of a Scientific Use File for the IAB-Establishment-Panel, Statistical Data Confidentiality, Proceedings of the Joint Eurostat/UN-ECE Work session on Statistical Data Confidentiality in March 1999.
3. Chauduri, A. and R. Mukerjee (1988): Randomized Response: Theory and Techniques, New York, Base: Marcel Dekker.
4. de Waal, A.G. and L.C.R.J. Willenborg (1997): Statistical Disclosure Control and Sampling Weights, Journal of Official Statistics 13, 417–434.
5. Domingo-Ferrer, J. and J.M. Mateo-Sanz (2001): Comparing SDC Methods for Microdata on the Basis of Information Loss and Disclosure Risk, paper presented at the Joint ECE/Eurostat Worksession on Statistical Confidentiality in Skopie (The former Yugoslav Republic of Macedonia), 14-16 March 2001
6. Fienberg, S.E. (1997): Confidentiality and Disclosure Limitation Methodology: Challenges for National Statistics and Sstatistical Research, Technical Report No. 611, Carnegie Mellon university, Pittsburgh.
7. Fuller, W.A. (1993): Masking Procedures for Microdata Disclosure Limitation, Journal of Official Statistics 9, 383–406.
8. Gouweleeuw, J.M., Kooiman, P., Willenborg, L.C.R.J. and P.-P. de Wolf (1998): Post Randomisation for Statistical Disclosure Control: Theory and Implementation, Journal of Official Statistics 14, 463–478.
9. Harhoff, D. and G. Licht (1993): Anonymisation of Innovation Survey Data, the Disguise Factor Method, Zentrum für Europäische Wirtschaftsforschung, Mannheim.
10. Kim, J.J. (1986): A Method for Limiting Disclosure in Microdata based on Random Noise and Transformation, Proceedings of the Section on Survey Research Methods 1986, American Statistical Association, 303–308.
11. Kim, J.J. (1990): Subpopulation Estimation for the Masked Data, Proceedings of the Section on Survey Research Methods 1990, American Statistical Association, 456–461.
12. Kim, J.J. and W.E. Winkler (1995): Masking Microdata Files, Proceedings of the Section on Survey Research Methods 1995, American Statistical Association, 114–119.
13. Kim, J.J. and W.E. Winkler (1997): Masking Microdata Files, Statistical Research Division RR97/03, US Bureau of the Census, Washington, DC.
14. Kim, J.J. and W.E. Winkler (2001): Multiplicative Noise for Masking Continouus Data, unpublished manuscript.
15. McGuckin, R. H. and S.V. Nguyen (1990), Public Use Microdata: Disclosure and Usefulness, Journal of Economic and Social Development 16, 19–39.
16. McGuckin, R. H. (1993): Analytic Use of Economic Microdata: A Model for Researcher Access with Confidentiality Protection, Proceedings of the International Seminar on Statistical Confidentiality, 08. – 10. Sept. 1992, Dublin, Ireland; Eurostat, 83–97.
17. Moore, R.A. (1996): Analysis of the Kim-Winkler Algorithm for Masking Microdata files – How Much Masking is Necessary and Sufficient? Conjectures for the Development of a Controllable Algorithm, Statistical Research Division RR96/05, US Bureau of the Census, Washington D.C.

18. Paass, G. (1988): Disclosure Risk and Disclosure Avoidance for Microdata, Journal of Business & Economic Statistics 6, 487–500.
19. Roque, G. M. (2000): Masking Microdata Files with Mixtures of Multivariate Normal Distributions, unpublished PhD-Thesis, University of California, Riverside.
20. Spruill, N.L (1983) The Confidentiality and Analytic Usefulness of Masked Business Microdata, Proceedings of the Section on Survey Research Methods 1983, American Statistical Association, 602–610.
21. Sullivan, G.R. (1989): The Use of Added Error to Avoid Disclosure in Microdata Releases, unpublished PhD-Thesis, Iowa State University.
22. Sullivan, G.R. and W.A. Fuller (1989): The Use of Measurement Error to Avoid Disclosure, Proceedings of the Section on Survey Research Methods 1989, American Statistical Association, 802–807.
23. Sullivan, G.R. and W.A. Fuller (1990): Construction of Masking Error for Categorial Variables, Proceedings of the Section on Survey Research Methods 1990, American Statistical Association, 453–439.
24. Tendick, P. (1988): Bias Avoidance and Measures of Confidentiality for the Noise Addition Method of Database Disclosure Control, unpublished PhD-Thesis, University of California, Davis.
25. Tendick, P. (1991): Optimal Noise Addition for Preserving Confidentiality in Multivariate Data, Journal of Statistical Planning and Inference 27, 341–353.
26. Tendick, P. and N. Matloff (1994): A Modified Random Perturbation Method for Databse Security, ACM Transactions on Database Systems 19, 47–63.
27. Willenborg, L. and T. de Waal (1996): Statistical Disclosure Control in Practice, New York, Berlin, Heidelberg: Springer.
28. Winkler, W.E. (1995): Matching and Record Linkage, Cox, B.G., Binder, D.A., Chinnappa, B.N., Christianson, A., Colledge, M.J. und P.S. Kott (ed.): Business Survey Methods, Wiley Series in Probability and Mathematical Statistics, New York: John Wiley & Sons, 335 - 384.
29. Winkler, W.E. (1998): Re-identification Methods for Evaluating the Confidentiality of Analytically Valid Microdata, Statistical Data Protection 1998, Lisbon, Portugal.

Annex

Table 1. Simulated (original) data: Descriptive statistics

Var	mean	stdc	min	max
$VAR12$	70.838	191.929	-4.799	3112.978
$VAR2$	19.170	118.891	-1028.698	2667.735
$VAR32$	0.002	0.059	0	1.705
$VAR4$	932.699	2738.207	0.037	48349.171
$VAR41$	60.679	324.991	-214.547	4824.280
$VAR413$	1.007	16.091	0	464.986
$VAR414$	43.172	253.979	-214.646	4824.280

Table 2. Simulated (original) data: Correlations

$VAR12$	1	·	·	·	·	·	·
$VAR2$	0.168	1	·	·	·	·	·
$VAR32$	-0.005	-0.007	1	·	·	·	·
$VAR4$	0.608	0.134	-0.013	1	·	·	·
$VAR41$	0.370	0.119	-0.008	0.827	1	·	·
$VAR413$	0.052	-0.007	-0.003	0.028	0.074	1	·
$VAR414$	0.230	0.158	-0.007	0.676	0.877	0.033	1

Table 3. Masked data ($\alpha = 0.5$): Descriptive statistics

VAR	corr. noise		white noise		Kim		Sullivan	
	mean	stdc	mean	stdc	mean	stdc	mean	stdc
$VAR12$	70.838	194.923	70.838	235.404	70.838	161.328	72.356	209.938
$VAR2$	19.170	125.248	19.170	141.953	19.170	101.291	20.573	143.215
$VAR32$	0.002	0.0712	0.002	0.0740	0.002	0.0578	0.003	0.227
$VAR4$	932.699	3333.567	932.699	3422.842	932.699	2808.140	940.004	2916.022
$VAR41$	60.679	398.890	60.679	395.217	60.679	325.474	35.229	343.754
$VAR413$	1.007	19.2409	1.007	19.550	1.007	16.208	1.197	19.976
$VAR414$	43.172	406.615	43.172	308.845	43.172	323.277	17.861	287.521

Table 4. Masked data($\alpha = 0.5$): Descriptive statistics (minimum, maximum)

VAR	corr. noise		white noise		Kim		Sullivan	
	max	min	max	min	max	min	max	min
$VAR12$	3055.518	-138.847	3100.782	-394.219	2539.968	-81.345	3685.115	-5.179
$VAR2$	2696.091	-1009.537	2535.178	-1024.843	2193.605	-784.914	3242.880	-1111.011
$VAR32$	1.743	-0.127	1.775	-0.137	1.439	-0.109	2.181	-0.368
$VAR4$	47867.22	-4887.340	51906.347	-5227.395	41739.254	-4284.085	58061.244	0.020
$VAR41$	4520.781	-966.465	4827.800	-885.363	4087.285	-596.456	5044.503	-321.820
$VAR413$	455.695	-38.541	453.384	-32.447	379.517	-30.803	581.730	-0.001
$VAR414$	4992.814	-910.681	4770.511	-536.764	4065.321	-781.466	5305.704	-321.968

Table 5. Original and masked data ($\alpha = 0.5$): Number of zeros and number of observations between -1 and 1

VAR	original data		masked data			
			corr. noise	white noise	Kim	Sullivan
	0	$-1 < x < 1$	$-1 < x < 1$	$-1 < x < 1$	$-1 < x < 1$	$-1 < x < 1$
$VAR12$	166	241	9	6	13	240
$VAR2$	136	248	13	6	21	247
$VAR32$	998	999	999	999	999	998
$VAR4$	0	9	0	0	0	9
$VAR41$	492	621	6	2	2	133
$VAR413$	973	982	72	67	70	981
$VAR414$	513	650	1	9	3	141

Table 6. Masked data (correlated noise): Difference in correlations to original data, $\alpha = 0.5$

$VAR12$	0	·	·	·	·	·	·
$VAR2$	-0.0044	0	·	·	·	·	·
$VAR32$	0.0008	0.01273	-2.22e-016	·	·	·	·
$VAR4$	-0.1112	-0.0229	0.0058	-4.44e-016	·	·	·
$VAR41$	-0.0678	-0.0007	-0.0027	-0.2670	1.11e-016	·	·
$VAR413$	-0.0206	0.01891	0.0263	-0.0403	-0.0406	2.22e-016	·
$VAR414$	-0.0743	-0.0793	0.0011	-0.3067	-0.4085	-0.0641	3.33e-016

Table 7. Masked data (white noise): Difference in correlations to original data, $\alpha = 0.5$

```
VAR12        0          ·         ·        ·       ·       ·       ·
VAR2      0.0003 2.22e-016         ·        ·       ·       ·       ·
VAR32  -0.0375    -0.0178       0        ·       ·       ·       ·
VAR4   -0.0011    -0.0008 -0.0247 -3.33e-016         ·       ·       ·
VAR41  -0.0262     0.0117 -0.0118   -0.0469       0       ·       ·
VAR413 -0.1123    -0.0149  0.0004   -0.0219 -0.0805       0       ·
VAR414 -0.1815    -0.1271  0.0071   -0.5741 -0.7590 -0.0343 3.33e-016
```

Table 8. Masked data (Kim's algorithm): Difference in correlations to original data, $\alpha = 0.5$

```
VAR12  1.11e-016         ·         ·         ·        ·        ·          ·
VAR2     0.0286       0         ·         ·        ·        ·          ·
VAR32    0.0123 0.0140 1.11e-016          ·        ·        ·          ·
VAR4     0.1073 0.0343   -0.0079 2.22e-016         ·        ·          ·
VAR41    0.0741 0.0342   -0.0045    0.2814 -3.33e-016 ·          ·
VAR413   0.0155 0.0135   -0.0217    0.0476     0.0713 2.22e-016          ·
VAR414   0.0949 0.0841   -0.0292    0.3033     0.4147    0.0271 -1.11e-016
```

Table 9. Masked data (Sullivan's algorithm): Difference in correlations to original data ($\alpha = 0.5$)

```
VAR12        0          ·          ·         ·        ·       ·       ·
VAR2   -0.0081 -2.2e-016          ·         ·        ·       ·       ·
VAR32   0.0025    -0.017        0         ·        ·       ·       ·
VAR4   -0.0020    0.0039 -0.00021 -2.2e-016         ·       ·       ·
VAR41  -0.0078    0.0018  0.00046   -0.031       0       ·       ·
VAR413 -0.0011   0.00030  -0.0024   0.0046 0.0054 2.2e-016          ·
VAR414  0.0027   -0.0019   -0.013   -0.025 0.0077   -0.014 1.1e-016
```

Table 10. Masked data ($\alpha = 0.5$): Number of false pairs with higher matching probability than the true pairs

Number of false pairs	corr. noise	white noise	Kim	Sullivan
≥ 6	863	928	863	976
5	11	4	6	1
4	12	7	10	4
3	14	5	10	5
2	17	8	22	4
1	21	10	27	3
0	62	38	62	7

Sensitive Micro Data Protection Using Latin Hypercube Sampling Technique

Ramesh A. Dandekar[1], Michael Cohen[2], and Nancy Kirkendall[1]

[1] Energy Information Administration, U. S. Department of Energy, Washington DC
{Ramesh.Dandekar,Nancy.Kirkendall}@eia.doe.gov
[2] National Academy of Sciences, Washington DC
MCohen@NAS.edu

Abstract. We propose use of Latin Hypercube Sampling to create a synthetic data set that reproduces many of the essential features of an original data set while providing disclosure protection. The synthetic micro data can also be used to create either additive or multiplicative noise which when merged with the original data can provide disclosure protection. The technique can also be used to create hybrid micro data sets containing pre-determined mixtures of real and synthetic data. We demonstrate the basic properties of the synthetic data approach by applying the Latin Hypercube Sampling technique to a database supported a by the Energy Information Administration. The use of Latin Hypercube Sampling, along with the goal of reproducing the rank correlation structure instead of the Pearson correlation structure, has not been previously applied to the disclosure protection problem. Given its properties, this technique offers multiple alternatives to current methods for providing disclosure protection for large data sets.

1 Introduction

The utility of public-use microdata, generated either synthetically or through use of other techniques, is assessed by how well inferences based on this modified data set mimic those derived through analysis of the real data. This assessment naturally includes univariate as well as multivariate analyzes. We use Latin Hypercube Sampling of McKay et al. (1979) along with the rank correlation refinement of Iman and Conover (1982) to reproduce both the univariate and multivariate structure (in the sense of rank correlation) of the original data set. In doing this, much of the univariate and multivariate structure of the real data is retained. At the same time, the data is sufficiently altered to provide considerable protection from disclosure.

In the following, we first describe the basic procedure. We then discuss some refinements including smoothing of the empirical cumulative distribution function and application of the technique to subpopulations. We also present a real data application that helps to suggest some general advantages of the approach.

J. Domingo-Ferrer (Ed.): Inference Control in Statistical Databases, LNCS 2316, pp. 117–125, 2002.

2   LHS and the Restricted Pairing Algorithm

Univariate statistics of interest are generally simple functions of the first few moments such as the mean, standard deviation, and coefficient of skewness. While interest usually resides in these specific aspects of a distribution, a comprehensive picture of the univariate properties of any variable can be represented by the cumulative distribution function (cdf). Latin Hypercube Sampling (LHS), developed by McKay, et al. (1979), can be used to generate a synthetic data set for a group of uncorrelated variables in which the univariate characteristics of the original data are reproduced almost exactly. In the case where the variables are not uncorrelated, the restricted pairing algorithm of Iman and Conover (1982), which makes use of LHS, has the goal of producing a synthetic data set that reproduces the rank correlation structure of the real data, which has advantages and disadvantages in comparison with procedures that attempt to reproduce the Pearson correlation structure. One benefit of rank correlation is that for non-normal data, the rank correlation is a more useful summary of the relatedness of two variables that are monotonically but not linearly related. This approach would also have an advantage for heavily skewed distributions, for which the Pearson correlation can be dominated by a small percentage of the data. Finally, while the full procedure that we are advancing is slightly oriented to treatment of continuous variables, the treatment of discrete or categorical data is straightforward and is discussed below

The basic LHS procedure, in combination with the restricted pairing algorithm, proceeds as follows. Let V be an $n \times m$ data matrix containing n samples of $i/(n+1)$, $i = 1, \ldots, n$, for m different variables, $j = 1, \ldots, m$. Thus, initially V consists of m identical columns. To induce a desired rank correlation matrix for the variables in V, proceed as follows:

1. Let T be the $m \times m$ target rank correlation matrix we wish to impose on data matrix V.
2. Shuffle the columns of matrix V to remove perfect correlation to get V^*.[1]
3. Let C a be the $m \times m$ correlationtrix for data matrix V^* imposed by the shuffling.
4. Compute the $m \times m$ lower triangular matrix Q, such that $QQ' = C$, where $'$ denotes transpose.
5. Compute the $m \times m$ matrix P, such that $PP' = T$.
6. Compute the $m \times m$ matrix S, such that $S = PQ^{-1}$, where Q^{-1} is the inverse of matrix Q. (This creates S which operates by "removing" the correlation structure of C and instituting the correlation structure of T.)
7. Compute the $n \times m$ matrix R, such that $R = (V^*)S'$. (Note that R has a correlation structure that approximates T.)
8. By columns, replace the values of R with their corresponding ranks ranging from 1 to n.

---

[1] The algorithm yields slightly different rank correlation matrices depending on the starting values of the shuffled matrix V.

9. By columns, arrange the values of matrix V^* in the same rank sequence as the corresponding columns of R, to arrive at V^{**}.

10. Finally, for each column, replace each rank by the inverse cdf for that variable evaluated at $i/(n+1)$ to arrive at the modified data matrix. denumerate The resultant matrix has a rank correlation structure that resembles T.

3 Iterative Refinement

For a population with a large number of observations and/or a large number of variables, it is essential to mimic the multivariate covariance structure to the extent possible. With a large number of observations, correlations are known to great accuracy, and if the procedure introduces a comparable amount of error, the value of the synthetic data is reduced in comparison to use of the original data set. With a large number of variables, the procedure as outlined above tends to get further away from the target correlation matrix for individual correlations. Since microdata users could potentially be interested in any subset of the variables, reproducing each of the rank correlations gains importance. Iterative refinement of the rank correlation matrix, developed by Dandekar (1993), achieves that objective by reducing the gap between the rank correlations of the actual and the synthetic data.

Iterative refinement of the rank correlation matrix is based on the fact that the original algorithm yields slightly different results depending on the starting values of the shuffled matrix V^*. Iterative refinement is implemented as follows:

- Perform all the steps of the restricted pairing algorithm as described above.
- Let C^* be the revised $m \times m$ correlation matrix of data matrix V^{**} in the procedure above.
- Repeat steps 4 through 9 of the restricted pairing algorithm described above until the Mean Absolute Deviation (MAD) between the off-diagonal elements of T and updates of C^* cannot be reduced further within an acceptable tolerance level.

It has been observed that highly correlated variables tend to benefit the most from this iterative refinement procedure.

4 Choice of Cumulative Distribution Function

The cumulative distribution function (cdf) required to generate synthetic variables (step 10 above) could be constructed using either a fitted theoretical distribution or by using the observed empirical cumulative distribution function. The empirical cdf has the advantage of making no assumptions. However, it provides little information for the extreme tails of the distribution. Fitted theoretical cdfs are dependent on the assumed distributional family, but if the assumption is relatively correct, the resulting cdf might produce better values in the tails of the distribution.

To better control the univariate statistical properties of the synthetic data,4a, we prove using the empirical cdf for each of the population variables. Constructing the empirical cdf is straight-forward. We wish to be flexible in the algorithm to produce synthetic data sets of a different size than the original data set. Therefore, if we have n s for the original data set, we may be interested in producing a synthetic data set with N rows. Particularly if N is larger than n, some extrapolation will be needed in using the empirical cdf.

We make one other modification to the empirical cdf. To protect the true values for extreme outliers in the database,3e, we smooth the tails of the distribution as follows:

1. Choose fractional value for p, and choose some non-zero tolerance T.
2. If

$$\frac{\left|Y_{(j)} - Y_{(j-1)}\right|}{Y_{(j)}} > T,$$

replace any order statistic $Y_{(j)}$ with

$$pY(j) + (1 - p)Y(j - 1).$$

5   Treatment of Identifiable Subpopulations

When categorical variables identify subgroups that have clearly distinct distributions, either distinct rank correlations or univariate characteristics that distinguish these groups, a synthetic data set that can retain this structure will be more useful to the analyst. To accomplish this, a separate LHS sample can be drawn for subgroups that have specific values for combinations of these variables. In our example described below, the classification variables that might be examined for this include industry type, geological descriptors, income categories and many other categorical variables. By separately applying the procedure described here to subgroups, each subgroup in the synthetic data set can have its own statistical characteristics, since the LHS-based synthetic data will reproduce the statistical characteristics of each one of the subgroups identified in the data.

To be able to measure the rank correlations of each subgroup of the population, it is essential that each subgroup contain enough observations. When this basic condition is not satisfied, the subgroup must be merged into a larger unit by combining it with other closely related subgroup(s) until the rank correlation structure of the resultant subgroup can be determined to a needed degree of precision. Depending upon the number of observations, the subgroup collapsing procedure may need to be repeated over different combinations of classification variables until there are enough observations to determine the correlations.

When the subgroups are not identifiable, a synthetic data set can be generated by drawing a single LHS-based sample to represent the entire population. In this case, the categorical variables are simply additional variables which have a discrete univariate distribution where the cdf is a step function.

6   Size of the Synthetic Data Set

An additional useful feature of the LHS procedure proposed here is that it permits flexibility in the size of the synthetic data set that is generated. This is an important advantage, since it can be used to provide additional protection against the disclosure of outliers. These so-called "population uniques" generally appear at the tail end of the distributions, and can be used to identify specific members of the data set. By generating a synthetic data set several times larger than the original, and given the smoothing mentioned previously, the individual outlier becomes one of several observations that have moderately different values blurring the distinctiveness of these observations. The larger sample size also permits the distribution of the 'unique' original observation over different combinations of classification variables included in a subset of the data. This feature would be especially helpful for protection for establishment surveys. Such an approach prevents one to one comparisons, and at the same time helps retain the closeness to the overall statistical characteristics of the original data.

The challenge with simulating a larger sample size than we have observed is that in theory there should an increased chance of selecting an observation larger than the maximum of the observed sample. When the empirical sample cdf is used as the basis of the LHS sample, the maximum in the observed sample will determine the maximum in the simulated sample. How best to simulate observations in the tails of the distribution is a topic for future research.

On the other hand, by generating a synthetic data set several times smaller than the original, provides the capability to create micro aggregates of the original records. Micro aggregation is one of the techniques proposed to be used to protect micro data containing sensitive information. Unlike the conventional micro aggregates which are divided along the physical boundaries, the LHS-based micro-aggregates are partitioned using equi-probabilistc boundaries.

7   An Illustrative Example[2]

To demonstrate the LHS-based synthetic data generation procedure, we have used 13 variables from the Commercial Building Energy Consumption Survey (CBECS) carried out by the Energy Information Administration (EIA) of the U.S. Department of Energy.[3] Our test data set consists of two categorical variables (PBA and YRCON) and 11 continuous variables. The variables

---

[2] The following example involves data from a sample survey. While these typically have sample weights determined by a sample design that might involve stratification or clustering, for the current paper we are restricting our attention to surveys in which each data case has identical sampling weight. We hope to address this complication in future work.

[3] The Commercial Buildings Energy Consumption Survey is a national-level sample survey of commercial buildings and their energy suppliers conducted quadrennially by the Energy Information Administration.

selected for the analysis are: (1) electricity consumption—ELBTU, (2) electricity expenditures—ELEXP, (3) natural gas consumption—NGBTU, (4) natural gas expenditures—NGEXP, (5) major fuel consumption—MFBTU, (6) major fuel expenditures—MFEXP, (7) principal building activity—PBA, (8) year constructed—YRCON, (9) total floor space—SQFT, (10) number of employees—NWKER, (11) percent of floor space cooled—COOLP, (12) percent of floor space heated—HEATP, and (13) percent of floor space lit—LTOHRP.

Our test population consists of 5655 observations. The rank correlation matrix is as follows:

$$
\begin{pmatrix}
1.0000 & .9782 & .3658 & .3715 & .9273 & .9627 & -.0930 & .2313 & .8143 & .7650 & .3341 & .2236 & .2535 \\
.9782 & 1.0000 & .3652 & .3755 & .9113 & .9797 & -.0968 & .2252 & .8118 & .7640 & .3169 & .2076 & .2399 \\
.3658 & .3652 & 1.0000 & .9956 & .5159 & .4295 & .0817 & -.0232 & .3675 & .3153 & .0929 & .2439 & .0797 \\
.3715 & .3755 & .9956 & 1.0000 & .5173 & .4395 & .0782 & -.0183 & .3735 & .3224 & .0926 & .2374 & .0805 \\
.9273 & .9113 & .5159 & .5173 & 1.0000 & .9640 & -.0584 & .1263 & .8341 & .7532 & .2447 & .2891 & .2218 \\
.9627 & .9797 & .4295 & .4395 & .9640 & 1.0000 & -.0846 & .1847 & .8368379 & .7728 & .2837 & .2411 & .2335 \\
-.0930 & -.0968 & .0817 & .0782 & -.0584 & -.0846 & 1.0000 & -.0333 & -.1409 & -.2159 & -.0330 & .0801 & .0769 \\
.2313 & .2252 & -.0232 & -.0183 & .1263 & .1847 & -.0333 & 1.0000 & .1199 & .1597 & .2364 & .0543 & .1583 \\
.8143 & .8118 & .3675 & .3735 & .8341 & .8368379 & -.1409 & .1199 & .0000 & .7605 & .1075 & .0942 & .1340 \\
.7650 & .7640 & .3153 & .3224 & .7532 & .7728 & -.2159 & .1597 & .7605 & 1.0000 & .2416 & .1782 & .2065 \\
.3341 & .3169 & .0929 & .0926 & .2447 & .2837 & -.0330 & .2364 & .1075 & .2416 & 1.0000 & .4213 & .2140 \\
.2236 & .2076 & .2439 & .2374 & .2891 & .2411 & .0801 & .0543 & .0942 & .1782 & .4213 & 1.0000 & .2553 \\
.2535 & .2399 & .0797 & .0805 & .2218 & .2335 & .0769 & .1583 & .1340 & .2065 & .2140 & .2553 & 1.0000
\end{pmatrix}
$$

By following the procedure outlined here, a total of 5,000 synthetic observations, consisting of 13 synthetic variables, were generated. The original data set was used to construct the cdfs. A smoothing constant $p = 0.5$ was used to smooth the ends of the cdfs. Each synthetic observation carries a sampling weight of 1.13 $(5,655/5,000)$. The sampling weight is required to match totals from the simulated data with totals from the original data.

A total of nine iterative refinement steps were required to bring the synthetic data rank correlation matrix to within 0.01% of the target rank correlation matrix. The rank correlation structure of the synthetic data matrix at the end of the initial step and at the end of the last refinement step, along with the sum absolute deviation of the upper (or lower) off-diagonal elements of the rank correlation matrix at the end of each iterative step are as follows:

Iteration: 0 — Abs Diff: $.7188911E + 00$

Rank Correlation Matrix:

$$
\begin{pmatrix}
1.0000 & .9794 & .3482 & .3534 & .9319 & .9649 & -.0891 & .2190 & .8203 & .7681 & .3158 & .2119 & .2387 \\
.9794 & 1.0000 & .3477 & .3570 & .9174 & .9809 & -.0924 & .2138 & .8180 & .7669 & .3004 & .1975 & .2266 \\
.3482 & .3477 & 1.0000 & .9959 & .4928 & .4097 & .0818 & -.0258 & .3520 & .3011 & .0844 & .2294 & .0707 \\
.3534 & .3570 & .9959 & 1.0000 & .4943 & .4187 & .0784 & -.0222 & .3577 & .3075 & .0833 & .2235 & .0716 \\
.9319 & .9174 & .4928 & .4943 & 1.0000 & .9663 & -.0582 & .1260 & .8386 & .7577 & .2365 & .2713 & .2113 \\
.9649 & .9809 & .4097 & .4187 & .9663 & 1.0000 & -.0821 & .1776 & .8419 & .7753 & .2708 & .2282 & .2216 \\
-.0891 & -.0924 & .0818 & .0784 & -.0582 & -.0821 & 1.0000 & -.0309 & -.1340 & -.2059 & -.0285 & .0754 & .0737 \\
.2190 & .2138 & -.0258 & -.0222 & .1260 & .1776 & -.0309 & 1.0000 & .1183 & .1515 & .2179 & .0500 & .1460 \\
.8203 & .8180 & .3520 & .3577 & .8386 & .8419 & -.1340 & .1183 & 1.0000 & .7608 & .1125 & .0956 & .1332 \\
.7681 & .7669 & .3011 & .3075 & .7577 & .7753 & -.2059 & .1515 & .7608 & 1.0000 & .2318 & .1685 & .1939 \\
.3158 & .3004 & .0844 & .0833 & .2365 & .2708 & -.0285 & .2179 & .1125 & .2318 & 1.0000 & .3981 & .1955 \\
.2119 & .1975 & .2294 & .2235 & .2713 & .2282 & .0754 & .0500 & .0956 & .1685 & .3981 & 1.0000 & .2366 \\
.2387 & .2266 & .0707 & .0716 & .2113 & .2216 & .0737 & .1460 & .1332 & .1939 & .1955 & .2366 & 1.0000
\end{pmatrix}
$$

Iteration: 1 — Abs Diff: $.6919879E - 01$
Iteration: 2 — Abs Diff: $.9697238E - 02$
Iteration: 3 — Abs Diff: $.3696279E - 02$
Iteration: 4 — Abs Diff: $.3167076E - 02$
Iteration: 5 — Abs Diff: $.3095646E - 02$

Iteration: 6 — Abs Diff: $.3083975E - 02$
Iteration: 7 — Abs Diff: $.3081255E - 02$
Iteration: 8 — Abs Diff: $.3080559E - 02$
Iteration: 9 — Abs Diff: $.3080209E - 02$

Rank Correlation Matrix:

```
1.0000   .9782   .3657   .3714   .9273   .9627  -.0930   .2313   .8143   .7650   .3340   .2236   .2534
 .9782  1.0000   .3651   .3754   .9113   .9797  -.0967   .2252   .8118   .7640   .3169   .2076   .2399
 .3657   .3651  1.0000   .9956   .5158   .4294   .0817  -.0232   .3674   .3152   .0929   .2439   .0797
 .3714   .3754   .9956  1.0000   .5172   .4394   .0782  -.0183   .3734   .3223   .0926   .2374   .0805
 .9273   .9113   .5158   .5172  1.0000   .9640  -.0584   .1263   .8341   .7532   .2447   .2891   .2218
 .9627   .9797   .4294   .4394   .9640  1.0000  -.0846   .1847   .8379   .7728   .2837   .2411   .2335
-.0930  -.0967   .0817   .0782  -.0584  -.0846  1.0000  -.0333  -.1409  -.2159  -.0330   .0801   .0769
 .2313   .2252  -.0232  -.0183   .1263   .1847  -.0333  1.0000   .1199   .1596   .2363   .0543   .1583
 .8143   .8118   .3674   .3734   .8341   .8379  -.1409   .1199  1.0000   .7605   .1075   .0942   .1340
 .7650   .7640   .3152   .3223   .7532   .7728  -.2159   .1596   .7605  1.0000   .2416   .1782   .2065
 .3340   .3169   .0929   .0926   .2447   .2837  -.0330   .2363   .1075   .2416  1.0000   .4212   .2139
 .2236   .2076   .2439   .2374   .2891   .2411   .0801   .0543   .0942   .1782   .4212  1.0000   .2552
 .2534   .2399   .0797   .0805   .2218   .2335   .0769   .1583   .1340   .2065   .2139   .2552  1.0000
```

The resulting rank correlation matrix is now almost identical to the target. The Pearson correlation coefficient, unlike the rank correlation coefficient, is extremely sensitive to the actual values of the variables and therefore is difficult to reproduce exactly. The Pearson rank correlation coefficient matrix for the original population and the synthetic data are as follows:

Pearson Correlation Coefficient - Actual Data

```
1.0000   .9292   .4212   .4436  .8068074   18  -.0412   .1086   .7241   .0277   .1816   .0926   .1131
 .9292  1.0000   .3624   .3917   .7457   .9824  -.0452   .1008   .7393   .0383   .1844   .0928   .1118
 .4212   .3624  1.0000   .9500   .7459   .4601  -.0244   .0105   .3373  -.0037   .1020   .0913   .0641
 .4436   .3917   .9500  1.0000   .7395   .4949   .0194   .0208   .3680  -.0062   .1079   .1045   .0728
 .8074   .7457   .7459   .7395  1.0000   .8203   .0152   .0601   .6330   .0130   .1669   .1165   .1052
 .9218   .9824   .4601   .4949   .8203  1.0000  -.0413   .0903   .7518   .0338   .1885   .1074   .1158
-.0412  -.0452   .0244   .0194   .0152  -.0413  1.0000  -.0183  -.0635  -.1084  -.0310   .0659   .0587
 .1086   .1008   .0105   .0208   .0601   .0903  -.0183  1.0000   .1090  -.0357   .2251   .0513   .1589
 .7241   .7393   .3373   .3680   .6330   .7518  -.0635   .1090  1.0000   .0479   .1392   .0798   .0968
 .0277   .0383  -.0037  -.0062   .0130   .0338  -.1084  -.0357   .0479  1.0000  -.0974  -.1345  -.3586
 .1816   .1844   .1020   .1079   .1669   .1885  -.0310   .2251   .1392  -.0974  1.0000   .4494   .2635
 .0926   .0928   .0913   .1045   .1165   .1074   .0659   .0513   .0798  -.1345   .4494  1.0000   .3288
 .1131   .1118   .0641   .0728   .1052   .1158   .0587   .1589   .0968  -.3586   .2635   .3288  1.0000
```

Pearson Correlation Coefficient - Synthetic Data

```
1.0000   .6883   .2485   .2366   .4731   .5835  -.0468   .1224   .4140   .1965   .1317   .0793   .1023
 .6883  1.0000   .2737   .2860   .4223   .7481  -.0573   .1175   .4293   .2592   .1219   .0716   .1008
 .2485   .2737  1.0000   .9335   .4557   .3638   .0135   .0192   .2461   .1047   .0529   .0544   .0460
 .2366   .2860   .9335  1.0000   .4077   .3683   .0143   .0259   .2435   .1312   .0629   .0689   .0500
 .4731   .4223   .4557   .4077  1.0000   .6804  -.0181   .0276   .5273   .2282   .0661   .1064   .0830
 .5835   .7481   .3638   .3683   .6804  1.0000  -.0465   .0796   .5440   .2988   .1028   .0934   .1014
-.0468  -.0573   .0135   .0143  -.0181  -.0465  1.0000  -.0483  -.1010  -.0834  -.0402   .0551   .0494
 .1224   .1175   .0192   .0259   .0276   .0796  -.0483  1.0000   .0558   .0335   .2053   .0417   .1557
 .4140   .4293   .2461   .2435   .5273   .5440  -.1010   .0558  1.0000   .3128   .0271   .0472   .0699
 .1965   .2592   .1047   .1312   .2282   .2988  -.0834   .0335   .3128  1.0000   .0328   .0339   .0384
 .1317   .1219   .0529   .0629   .0661   .1028  -.0402   .2053   .0271   .0328  1.0000   .3348   .1854
 .0793   .0716   .0544   .0689   .1064   .0934   .0551   .0417   .0472   .0339   .3348  1.0000   .2125
 .1023   .1008   .0460   .0500   .0830   .1014   .0494   .1557   .0699   .0384   .1854   .2125  1.0000
```

Our test population is highly skewed. To better understand the difference this has made on the Pearson correlations, we transformed our original and the synthetic data using logarithms and computed the Pearson correlation coefficient of the transformed data. The resulting Pearson correlation matrix now is almost identical to each other. This is especially true of the highly correlated variables.

Pearson Correlation of log(1+Variable) - Real Data

$$
\begin{pmatrix}
1.0000 & .9552 & .2369 & .3057 & .9018 & .9255 & -.0930 & .2290 & .7798 & .5946 & .3955 & .2866 & .3188 \\
.9552 & 1.0000 & .2440 & .3154 & .8791 & .9634 & -.1000 & .2221 & .7900 & .6172 & .3725 & .2604 & .3008 \\
.2369 & .2440 & 1.0000 & .9860 & .3930 & .3117 & .0984 & -.0596 & .2420 & .1326 & .1393 & .2803 & .1275 \\
.3057 & .3154 & .9860 & 1.0000 & .4636 & .3875 & .0874 & -.0351 & .3177 & .1940 & .1449 & .2790 & .1388 \\
.9018 & .8791 & .3930 & .4636 & 1.0000 & .9525 & -.0446 & .1283 & .8119 & .5969 & .3341 & .3853 & .3243 \\
.9255 & .9634 & .3117 & .3875 & .9525 & 1.0000 & -.0864 & .1831 & .8314 & .6378 & .3543 & .3107 & .3090 \\
-.0930 & -.1000 & .0984 & .0874 & -.0446 & -.0864 & 1.0000 & -.0303 & -.1403 & -.2612 & -.0272 & .1097 & .1110 \\
.2290 & .2221 & -.0596 & -.0351 & .1283 & .1831 & -.0303 & 1.0000 & .1285 & .1238 & .2143 & .0360 & .1334 \\
.7798 & .7900 & .2420 & .3177 & .8119 & .8314 & -.1403 & .1285 & 1.0000 & .6811 & .1589 & .1055 & .1380 \\
.5946 & .6172 & .1326 & .1940 & .5969 & .6378 & -.2612 & .1238 & .6811 & 1.0000 & .2064 & .1132 & -.0374 \\
.3955 & .3725 & .1393 & .1449 & .3341 & .3543 & -.0272 & .2143 & .1589 & .2064 & 1.0000 & .4489 & .2728 \\
.2866 & .2604 & .2803 & .2790 & .3853 & .3107 & .1097 & .0360 & .1055 & .1132 & .4489 & 1.0000 & .3445 \\
.3188 & .3008 & .1275 & .1388 & .3243 & .3090 & .1110 & .1334 & .1380 & -.0374 & .2728 & .3445 & 1.0000
\end{pmatrix}
$$

Pearson Correlation of log(1+Variable) - Synthetic Data

$$
\begin{pmatrix}
1.0000 & .9432 & .2975 & .3210 & .8716 & .9201 & -.0944 & .2379 & .7704 & .6404 & .2763 & .1795 & .2441 \\
.9432 & 1.0000 & .3038 & .3325 & .8616 & .9641 & -.1034 & .2318 & .7811 & .6579 & .2616 & .1621 & .2203 \\
.2975 & .3038 & 1.0000 & .9690 & .4686 & .3780 & .0789 & -.0382 & .3206 & .2415 & .0635 & .1790 & .0647 \\
.3210 & .3325 & .9690 & 1.0000 & .4916 & .4086 & .0805 & -.0283 & .3425 & .2617 & .0707 & .1781 & .0719 \\
.8716 & .8616 & .4686 & .4916 & 1.0000 & .9435 & -.0519 & .1135 & .8070 & .6389 & .1965 & .2384 & .2081 \\
.9201 & .9641 & .3780 & .4086 & .9435 & 1.0000 & -.0843 & .1826 & .8182 & .6742 & .2305 & .1951 & .2156 \\
-.0944 & -.1034 & .0789 & .0805 & -.0519 & -.0843 & 1.0000 & -.0417 & -.1448 & -.1951 & -.0383 & .0620 & .0644 \\
.2379 & .2318 & -.0382 & -.0283 & .1135 & .1826 & -.0417 & 1.0000 & .1217 & .1446 & .2021 & .0460 & .1532 \\
.7704 & .7811 & .3206 & .3425 & .8070 & .8182 & -.1448 & .1217 & 1.0000 & .6758 & .0840 & .0765 & .1233 \\
.6404 & .6579 & .2415 & .2617 & .6389 & .6742 & -.1951 & .1446 & .6758 & 1.0000 & .1684 & .1218 & .1639 \\
.2763 & .2616 & .0635 & .0707 & .1965 & .2305 & -.0383 & .2021 & .0840 & .1684 & 1.0000 & .3345 & .1967 \\
.1795 & .1621 & .1790 & .1781 & .2384 & .1951 & .0620 & .0460 & .0765 & .1218 & .3345 & 1.0000 & .2210 \\
.2441 & .2203 & .0647 & .0719 & .2081 & .2156 & .0644 & .1532 & .1233 & .1639 & .1967 & .2210 & 1.0000
\end{pmatrix}
$$

So far as the univariate statistical properties are concerned, the LHS-based procedure is well known to reproduce them almost exactly. This is apparent from Table below.

Table 1. Univariate Statistics

Var	Original Average	Population Std Dev	Synthetic Average	Data Std Dev
1	$.7929E + 07$	$.2312E + 08$	$.7961E + 07$	$.2345E + 08$
2	$.1539E + 06$	$.4348E + 06$	$.1544E + 06$	$.4394E + 06$
3	$.4498E + 07$	$.2109E + 08$	$.4516E + 07$	$.2126E + 08$
4	$.1546E + 05$	$.5900E + 05$	$.1551E + 05$	$.5943E + 05$
5	$.1523E + 08$	$.4570E + 08$	$.1528E + 08$	$.4622E + 08$
6	$.1848E + 06$	$.4956E + 06$	$.1853E + 06$	$.4999E + 06$
7	$.1238E + 02$	$.1089E + 02$	$.1238E + 02$	$.1089E + 02$
8	$.5398E + 01$	$.1799E + 01$	$.5399E + 01$	$.1798E + 01$
9	$.1293E + 06$	$.2671E + 06$	$.1294E + 06$	$.2674E + 06$
10	$.1385E + 04$	$.1068E + 05$	$.1390E + 04$	$.1068E + 05$
11	$.6492E + 02$	$.4055E + 02$	$.6493E + 02$	$.4055E + 02$
12	$.8412E + 02$	$.3123E + 02$	$.8412E + 02$	$.3123E + 02$
13	$.8828E + 02$	$.2504E + 02$	$.8828E + 02$	$.2502E + 02$

8   Real Life Implementation of LHS-Based Technique

The Latin Hypercube Sampling technique offers a viable alternative to other methods of protection of sensitive data in public use data files. The LHS based technique could be implemented in a variety of different ways to achieve different end results in sensitive micro data disclosure avoidance arena. As described in this paper, the technique could be used to create synthetic micro data

- In stand alone mode using all the variables. When the sample size is order of magnitude smaller than the original population size, micro aggregation along the probabilistic boundaries results. On the other extreme, when the sample size is far larger than the original population, it creates a super population in which blurring of original data occurs.
- By selecting only a few variables from the original population which are deemed highly sensitive and operating on those variables.

The other implementation possibilities include:

- Combining real database containing sensitive data with LHS based synthetic database in predetermined proportion to create hybrid data base.
- The LHS based synthetic data base could be combined with the original data base to induce (a) an additive noise

$$N = x * O + (1 - x) * H,$$

where x is a fraction, or (b) as a multiplicative noise

$$N = O^{(a-b)/a} * H^{b/a},$$

where $b < a$. In this implementation x, a and b could either be fixed or random variables.

References

1. Dandekar, Ramesh Ch.A. (1993), "Performance Improvement of Restricted Pairing Algorithm for Latin Hypercube Sampling", ASA Summer conference (unpublished).
2. Iman R.L. and Conover W. J. (1982), "A Distribution-Free Approach to Inducing Rank Correlation Among Input Variables", *Commun. Stat.,* B11(3): pp. 311-334.
3. McKay M.D., Conover W. J., and Beckman, R. J. (1979), "A Comparison of Three Methods for Selecting Values of Input Variables in the Analysis of Output from a Computer Code", *Technometrics* 21(2): pp. 239-245.
4. Stein M. (1987), "Large Sample Properties of Simulations Using Latin Hypercube Sampling", *Technometrics* (29)2: pp. 143-151.

Integrating File and Record Level Disclosure Risk Assessment*

Mark Elliot

Centre for Census and Survey Research, University of Manchester
M13 9PL, United Kingdom

Abstract. This paper examines two methods of record-level disclosure risk assessment for microdata. The first uses an extension of the Special Uniques Identification method [7] combined with data mining techniques and the second uses Data Intrusion Simulation methodology [4,14] at the record level. Some numerical studies are presented showing the value of these two methods and proposals for integrating them with file level measures in risk driven file construction system are presented.
Keywords: Disclosure Risk Assessment, Microdata, Data Intrusion Simulation, Special Uniques.

1 Introduction

The accurate and reliable assessment of statistical disclosure risk is a necessary pre-cursor of the efficient application of disclosure control techniques. Research on statistical disclosure risk assessment with respect to microdata files has two major themes. The first uses the concept of uniqueness, examining the level of unique individuals or households within populations and samples [2,1,10,13]. The second attempts to model actual disclosure attempts by matching individual records in a target file with those in identification file (for example [12]. Both of these approaches are problematic. The uniqueness statistics require population data to calculate them and do not relate directly to what a data intruder might actually do in attempting to identify an individual within a dataset. The matching experiment approaches have a direct relationship but are ad hoc with respect to the identification data set chosen.

Elliot [4] has developed a method called *Data Intrusion Simulation* (DIS) for calculating the general risk for a given target file, which does not require population statistics or matching experiments and is grounded in the matching process that the intruder must use in order to identify individuals within an anonymised dataset. Elliot presents a numerical study demonstrating that the method gives accurate estimates of matching metrics such as the probability of a correct match given a unique match and therefore is a good measure of real disclosure risk. Skinner and Elliot [14] provide a statistical proof that the method does produce unbiased estimates of the population level of the probability of a

---

* The work described in this paper was supported by the UK Economic and Social Research Council, grant number R000 22 2852.

J. Domingo-Ferrer (Ed.): Inference Control in Statistical Databases, LNCS 2316, pp. 126–134, 2002.
© Springer-Verlag Berlin Heidelberg 2002

correct match given a unique match. This is a breakthrough in that it is the first time that a population level risk metric has been accurately estimated from sample data.

The remainder is divided into four parts the first describes the DIS method. The second shows how it might be extended to the record level of risk. The third presents a numerical demonstration of the extended method and the fourth discusses how the different levels of analysis might be integrated.

2   Data Intrusion Simulation (DIS)

The DIS method has similarities of form to the bootstrapping methods [3]. The basic principle of the method is to remove a small number records from the target microdata file and then copy back some of those records, with each record having a probability of being copied back equal to the sampling fraction of the original microdata file. This creates two files, a new slightly truncated target file and a file of the removed records, which is then matched against the target file. The method has two computational forms, the *special form*, where the sampling, described above, is actually done and the *general form*, where the sampling is not done but the equivalent effect is derived using the partition structure[1].

2.1   The Special Method

The special method follows the following five-step procedure, (a schematic version can be found in Appendix B).

1. Take a sample microdata file (A) with sampling fraction f.
2. Remove a small random number of records (B) from A, to make a new file (A').
3. Copy back a random number of the records from B to A' with each record having a probability of being copied back equal to f. The result of this procedure is that B will now represent a fragment of an outside database (an identification file) with an overlap with the A' equivalent to that between the microdata file and an arbitrary identification file with zero data divergence (with no identical values on the matching keys for the same individual).
4. Match B against A'. Generate an estimate of the matching metrics particularly, the probability of a correct match given a unique match, $pr(cm \setminus um)$, between the fragment B and the file A'.
5. Iterate through stages i-iv until the estimate stabilises.

---

[1] The partition structure of a file is a frequency of frequencies for a given cross-classification. A partition class is a particular cell within the partition structure. So, for example, *uniques* is the partition class where the sample frequency is 1. This has also been referred to as the *equivalence class structure*; [11] and sampling fraction of the microdata file.

2.2   DIS: The General Method

A more general method can be derived from the above procedure. Imagine that the removed fragment (B) is just a single record. There are six possible outcomes depending on whether the record is copied back or not and whether it was a unique, in a pair or in a larger partition class.

Table 1. Possible per record outcome from the DIS general method

record is	Copied back	not copied back
sample unique	**correct unique match**	non-match
one of a sample pair	multiple match including correct	**false unique match**
one of a larger equivalence class	multiple match including correct	false multiple match

The two critical cells in the above table are: where a unique record is copied back - this creates a correct unique match where one of a sample pair is not. - this creates a false unique match.

Given this, we can induce that the relative numbers in these two cells determine the probability of a correct match given a unique match; $pr(cm \setminus um)$. Given this, it is possible to shortcut the special method[2],since one can derive an estimated probability of a correct match given a unique, match from:

$$pr(cm \setminus um) \cong \frac{U \times f}{U \times f + P \times (1 - f)}$$

Where U is the number of sample uniques, P is the number of records in pairs and f is the sampling fraction.

So the general technique provides a simple method for estimating the probability of an intruder being able to match an arbitrary outside file to a given target microdata file. Numerical studies [4,14] have demonstrated empirically that the above method is sound at the file level of risk, in that it provides stable estimates of the population level of matching probabilities. Further, Skinner and Elliot provide proof that $pr(cm \setminus um)$ estimated using the general method is an unbiased estimator for the real matching probability. [3]

However, as the method stands the DIS provides only a *file level* measure of disclosure risk. This has been very useful in for example comparing a proposed data file with an existing one [16] or examining the impacts of disclosure control techniques [5] Recently, it has been recognised that risk varies across any given data file and so also needs to be analysed at levels other than the whole file [4,15,10] and in particular at the level of individual records.

---

[2] As later discussion will show the special method is still necessary to, for example, assess the impact of a disclosure control method that does not systematically alter the partition structure of a file.

[3] See [14] for an a method of estimating the error of the estimates.

3   Extending DIS to the Record Level

It is quite possible for DIS to work at levels below the whole file level. Table 2 (from [4]) shows how risk might vary depending upon the value for a particular key variable.

Table 2. $pr(cm \setminus um)$ for the key (age(94), sex(2) and marital status (5) on the 2% Individual Samples of Anonymised Records broken down by marital status

Variable value	$Pr((cm \setminus um)$—Value)
Marital Satus	$Pr(cm \setminus um)$
Single	0.023
Married	0.019
Remarried	0.033
Divorced	0.036
Widowed	0.027

So while in general we use DIS to estimate for a given set of key variables the mean matching probabilities, by simply breaking the file into sub-files according to the values for one of the key variables we can obtain a DIS probability estimate for each record given that it belongs to the sub-file. If we extend this to all of the variables within a key (as exemplified in Table 3 then we can see that, for each record, a set of risk probabilities will be estimated based on their membership of each of the sub-groups represented by the values on the key variables.

Table 3. Example record with probabilities of correct match given unique match conditional on each of the values within record

Variable	value	$Pr((cm \setminus um)$—Value)
Age	16	0.043
Sex	Female	0.032
Marital Status	Married	0.049
Mean	-	0.041

The question then arises as to how the values might be integrated to provide a measure of risk for the record. There are two immediately obvious possibilities the mean and the maximum. The main rationale for using the mean is that the average record level measures of risk will be the same as the file level measure. The maximum may provide a more useful way of identifying risky records and indeed the risk - causing variables within those records. As the main thrust of the current paper is to examine the potential for integrating file and record level risk the following numerical study demonstrates the method using the mean.

3.1 Numerical Demonstration

In order to demonstrate the method outlined above the following study was conducted:

Using population data available to us, a file of one area of the UK with population of 450,000 was used. The following key variables were employed:

a94 Age(94 categories)
sx2 Sex(2)
ms5 Marital Status(5)
en10 Ethnicity(10)
pe11 Primary Economic Status(11)
cob42 Country of Birth(42)

The choice of this particular cross classification was partly constrained by the variables for which population data was available and then selected from those on the basis of scenario analyses conducted [6][4]. In fact many different cross-classifications have been tested, all of which yield similar results to those presented here.

Three levels of sampling fraction were tested: 2%, 5%, and 10%, using a geographically stratified sampling design. All parallel, geographically stratified samples of each sampling fraction were tested and the mean estimates of $pr(cm \setminus um)$ generated using the general form of DIS.

Using the population data it was possible to identify whether each record was population unique on the key variables a comparison of the mean $pr(cm \setminus um)$ values for population uniques and non-uniques allows us to establish whether the method picks out risky records. This is shown in Table 4, as it can be seen the population uniques had higher mean $pr(cm \setminus um)$ levels than non-population uniques.

These ratios indicate that spread over the entire file the effect is relatively mild, uniques do not show that much higher $pr(cm \setminus um)$ values than non-uniques. However Table 5 looks at the data from the other viewpoint. For each of five levels of $pr(cm \setminus um)$ - with bands of width 0.2, the probabilities of population uniqueness are recorded. That is, the table shows the proportion of records that are population unique, given a certain level of record level $pr(cm \setminus um)$. The results indicate that the higher the $pr(cm \setminus um)$ value the more likely the record is to be population unique.

The effect is quite substantial - records with $pr(cm \setminus um)$ values in the range 0.41-0.60 are more likely to be population unique than records in the 0-0.20 range by a factor of between 70 and 100 (with these data). It is interesting to

---

[4] In order to perform this analysis, ONS supplied anonymised 1991 Census data for seven UK Local Authorities to CCSR under contract. The data are kept in a secure environment, access is limited to members of the project team for the designated parts of the project, and the data will be returned to ONS when the work is complete. For the purpose of the contract, the CCSR is a supplier of services to the Registrar General for England and Wales and the 1991 Census Confidentiality Act applies.

Table 4. Mean $pr(cm \setminus um)$ per record for a key consisting of age, sex, marital status, economic status and ethnic group, conditional on each value within the key

Sampling fraction	Population Uniques	Non-uniques	ratio
2%	0.077	0.041	1.746
5%	0.100	0.078	1.293
10%	0.169	0.140	1.209

Table 5. The probability of population uniqueness by five bands of estimated per record $Pr(cm \setminus um)$ using the key age, sex, marital status, ethnic group, country of birth

	$Pr(cm \setminus um)$ - banded				
Sampling fraction	0-0.20	0.21-0.40	0.41-0.60	0.61-0.80	0.81-1.00
2%	0.002	0.007	0.095	0.175	-
5%	0.002	0.010	0.157	-	-
10%	0.001	0.007	0.147	-	-

note that this does not appear to interact with sampling fraction. That is for a given level of $pr(cm \setminus um)$ the probability of population uniqueness does not increase monotonically with sampling fraction.

4   The Special Uniques Method

The term special uniques was coined by Elliot [9] as a term to describe records within a dataset which were unique by virtue of processing a rare combination of characteristics as opposed to merely happening to be unique because of the sampling fraction and variable codings employed (random uniques). The technical definition first used for special uniques was a sample unique which maintained its uniqueness despite collapsing of the geographical component of its key variables.

Elliot and Manning [7] have shown that special uniques using this definition are far more likely to be population unique than are more likely to be population unique than random sample uniques. Elliot and Manning have also shown that the same principle could also be applied to variables other than geography and that persistence of uniqueness through aggregation of any variable indicated *specialness*. This in turn has lead to a re-definition of special uniqueness. In the final definition a special unique is: *a sample unique on key variable set K which is also sample unique on variable set k where k is a subset of K.*

Elliot and Manning [8] are currently developing a computational methodology to identify all uniques within a data set, and thereby classify records according degrees of specialness. It is envisaged that for each record within a dataset a *risk signature* could be generated, using all variables within a dataset. A records risk signature would indicate the number of unique variable pairs, triples, four folds and so on. By encoding these risk signatures within the dataset under a scenario based key variable generation scheme such as that proposed by Elliot and Dale (1999) it would be possible to provide a comprehensive risk analysis for a file both at the record and file levels which would allow differential treatment of records to optimise data quality for a given level of acceptable risk.

4.1   Numerical Study

In order to demonstrate the potential of this new method the following study was conducted:

The same population file, key variables and sampling fractions were used as in the DIS experiment reported above.

Each record within the dataset was given a score from 0 to 10 indicating how many sample unique variable pairs it was it had (using the five key variables there are 10 possible combinations of two variables). So, if there was one unique variable pair within a record then that record would be given a score of 1 and so on.

Table 6 shows the proportion of records with each number of sample unique variable pairs that are population unique on all five variables (age, sex, Marital status, primary economic status and ethnic group).

Table 6. The probability of population uniqueness given score of the number of sample unique variable pairs within the key age, sex. Marital status, primary economic status, and ethnic group

Score	Sampling Fraction				
	1%	2%	3%	5%	10%
0	0.001	0.001	0.001	0.002	0.003
1	0.031	0.075	0.115	0.164	0.268
2	0.254	0.417	0.493	0.582	0.700
3	0.475	0.634	0.800	0.739	1.000
4	0.226	0.143	0.000	1.000	-
5	0.667	-	-	-	-
6	-	-	-	-	-
7	-	-	-	-	-
8	-	-	-	-	-
9	-	-	-	-	-
10	-	-	-	-	-

The table clearly shows that special uniqueness within the scope of this new definition is highly predictive of population uniqueness. The larger the score on this simple measure the more likely the record is to be population unique. It should be stressed that this simple study only scratches the surface of what is possible. The study only uses five variables and does not consider larger unique variable combinations than pairs. Work is in progress to allow the total identification of the all sample uniques (of a number of variables) within a file and the development of more complex methods of classifying records based on their risk signatures, that is the pattern of sample uniqueness within each record.

5   Integrating Levels of Risk Analysis

Both the new special uniques method and the record level DIS method have demonstrable potential in terms of their ability to identify risky records. It is easy to see how with further development the DIS method could be extended to operate at multiple levels through a file. The special uniques method could be converted to a file level of measure of risk by considering the gross number of records exceeding a certain risk threshold across all scenarios deemed relevant for a particular file.

However, a better way forward would be to use the feature of the two methods in tandem at the file construction stage.

DIS could be used to evaluate the risk topology within a raw data file, with a view to vetting the proposed data structure at record level and also for identifying possibly risky variable combinations for including (bottom-up) in the scenario specifications to be used, a combined DIS/Special Uniques system could then identify the risky records within the proposed data set. These could then be manipulated/suppressed depending upon the severity of the risk, and then finally the gross file level risk could be checked using both DIS and gross special uniques counts, taking account of the impact disclosure control [5].

References

1. Bethlehem, J.G., Keller, W.J., and Pannekoek, J. Disclosure control of microdata, Journal of the American Statistical Association Vol. 85, 38-45, 1990.
2. Dalenius, T. Finding a Needle in a Haystack. *Journal of Official Statistics* Vol.2, No.3, 329–336, 1986.
3. Efron, B. Bootstrap methods: Another Look at the Jack-knife. *Annals of Statistics*, Vol. 7. 1–26, 1979.
4. Elliot, M.J. DIS: "A new approach to the measurement of statistical disclosure risk." *International Journal of Risk Management* 2(4) (2000): 39-48.
5. Elliot, M.J. "Data intrusion Simulation: Advances and a vision for the future of disclosure control." Paper presented to the 2nd UNECE work session on statistical data confidentiality; Skopje March 2001.
6. Elliot, M.J. and Dale, A. "Scenarios of Attack: The data intruder's perspective on statistical disclosure risk." *Netherlands Official Statistics.* Spring 1999.
7. Elliot, M.J., and Manning, A. "The Identification of Special Uniques". To appear in Proceedings Of GSS Methodology Conference. London. June 2001.
8. Elliot, M.J. and Manning, A. Statistical Disclosure Control and Data Mining. Proposal document to Economic and Social research Council under grant number (grant number R000 22 2852).

9. Elliot, M.J., Skinner, C.J., and Dale, A. "Special Uniques, Random Uniques and Sticky Populations: Some Counterintuitive Effects of Geographical Detail on Disclosure Risk". *Research in Official Statistics*; 1(2), 53–68, 1998.

10. Fienberg, S.E. and Makov, U.E., "Confidentiality Uniqueness and Disclosure Limitation for Categorical Data", *Journal of Official Statistics* 14(4), pp. 361–372, 1998.

11. Greenberg, B.V. and Zayatz, L.V. Strategies for Measuring Risk in Public Use Microdata Files. *Statistica Neerlandica*, 46, 33-48, 1992.

12. Muller, W., Blien, U., and Wirth, H. "Disclosure risks of anonymous individual data." Paper presented at the 1st International Seminar for Statistical Disclosure. Dublin, 1992.

13. Samuels, S.M. "A Bayesian, Species-Sampling-Inspired Approach to the Uniques Problem in Microdata Disclosure Risk Assessment." *Journal of Official Statistics*. 14(4) 373–383, 1998.

14. Skinner, C.J. and Elliot, M.J. 'A Measure of Disclosure Risk for Microdata'. CCSR occasional paper 23, 2001.

15. Skinner, C.J. and Holmes, D. J. , "Estimating the Re-identification Risk per Record", *Journal of Official Statistics* 14(4). pp. 361–372, 1998.

16. Tranmer, M., Fieldhouse E., Elliot, M.J., Dale A., and Brown, M. Proposals for Small Area Microdata. Accepted subject to revisions *Journal of the Royal Statistical Society*, Series A.

Disclosure Risk Assessment in Perturbative Microdata Protection*

William E. Yancey, William E. Winkler, and Robert H. Creecy

U.S. Bureau of the Census
{william.e.yancey,william.e.winkler,robert.h.creecy}@census.gov

Abstract. This paper describes methods for data perturbation that include rank swapping and additive noise. It also describes enhanced methods of re-identification using probabilistic record linkage. The empirical comparisons use variants of the framework for measuring information loss and re-identification risk that were introduced by Domingo-Ferrer and Mateo-Sanz.
Keywords: Additive noise, mixtures, rank swapping, EM Algorithm, record linkage.

1 Introduction

National Statistical Institutes (NSIs) have the need to provide public-use microdata that can be used for analyzes that approximately reproduce analyzes that could be performed on the non-public, original microdata. If microdata are analytically valid or have utility (see [18]), then re-identification of confidential information such as the names associated with some of the records may become easier.

This paper describes methods for masking microdata so that it is better protected against re-identification. The masking methods are rank swapping ([13], also [5]), additive noise ([8]), mixtures of additive noise ([14]). The re-identification methods are based on record linkage ([6], also [16]) with variants that are specially developed for re-identification experiments ([10], [18]). The overall framework is based on variants of methods that score both information loss and re-identification risk that were introduced by [5].

The outline of this paper is as follows. Section 2 covers the data files that were used in the experiments. In section 3, we describe the masking methods, the re-identification methods, and the variants of scoring metrics based on different types of information loss and re-identification risk. Section 4 provides results. In Section 5, we give discussion. The final section 6 consists of concluding remarks.

---

* This paper reports the results of research and analysis undertaken by Census Bureau staff. It has undergone a Census Bureau review more limited in scope than that given to official Census Bureau publications. This report is released to inform interested parties of research and to encourage discussion.

J. Domingo-Ferrer (Ed.): Inference Control in Statistical Databases, LNCS 2316, pp. 135–152, 2002.
© Springer-Verlag Berlin Heidelberg 2002

2   Data Files

Two data files were used.

2.1   Domingo-Ferrer and Mateo-Sanz

We used the same subset of American Housing Survey 1993 public-used data that was used by [5]. The Data Extraction System (http://www.census.gov/DES) was used to select 13 variables and 1080 records. No records having missing values or zeros were used.

2.2   Kim-Winkler

The original unmasked file of 59,315 records is obtained by matching IRS income data to a file of the 1991 March CPS data. The fields from the matched file originating in the IRS file are as follows:

1. Total income
2. Adjusted gross income
3. Wage and salary income
4. Taxable interest income
5. Dividend income
6. Rental income
7. Nontaxable interest income
8. Social security income
9. Return type
10. Number of child exemptions
11. Number of total exemptions
12. Aged exemption flag
13. Schedule D flag
14. Schedule E flag
15. Schedule C flag
16. Schedule F flag

The file also has match code and a variety of identifiers and data from the public-use CPS file. Because CPS quantitative data are already masked, we do not need to mask them. We do need to assure that the IRS quantitative data are sufficiently well masked so that they cannot easily be used in re-identifications either by themselves or when used with identifiers such as age, race, and sex that are not masked in the CPS file. Because the CPS file consists of a 1/1600 sample of the population, it is straightforward to minimize the chance of re-identification except in situations where a record may be a type of outlier in the population. For re-identification, we primarily need be concerned with higher income individuals or those with distinct characteristics that might be easily identified even when sampling rates are low.

3   Methods

The basic masking methods considered are (1) rank swapping, (2) univariate additive noise, and (3) mixtures of additive noise. Record linkage is the method of re-identification. The methods of information loss are those described by Domingo-Ferrer *et al.* [5]. and some variants.

3.1   Rank Swapping

Rank swapping was developed by Moore [13] and recently applied by Domingo-Ferrer *et al.* [5]. The data X is represented by (X_{ij}), $1 \leq i \leq n, 1 \leq j \leq k$, where i ranges through the number of records and j ranges through the number of variables. For each variable $j, 1 \leq j \leq k$, (X_{ij}) is sorted. For each j, (X_{ij}) can be swapped with (X_{il}) where $|j - l| < pn$ and p is a pre-specified proportion. The programming needed for implementing rank swapping is quite straightforward.

3.2   Additive Noise

Kim [8] introduced independent additive noise ε with covariance proportional to the original data X so that $Y = X + \varepsilon$ is the resultant masked data. The term ε has expected value 0. He showed that the covariance of Y is a multiple of the covariance of X and gave a transformation to another variable Z that is masked and has the same covariance as X. He also showed how regression coefficients could be computed and how estimates could be obtained on subdomains. His work has been extended by Fuller [7]. In this paper, we will consider the basic additive noise $Y = X + \varepsilon$ as was also considered by Fuller. Masking via additive noise has the key advantage that it can preserve means and covariances. Additive noise has the disadvantage that files may not be as confidential as with some of the other masking procedures. Kim [9] has shown that means and covariances from the original data can be reconstructed on all subdomains using the observed means and covariances from the masked data and a few additional parameters that the data provider must produce. Fuller [7] has shown that higher order moments such as the regression coefficients of interaction terms can be recovered provided that additional covariance information is available. In most situations, specialized software is needed for recovering estimates from the masked file that are very close to the estimates from the original, unmasked file.

3.3   Mixtures of Additive Noise

Roque [14] introduced a method of masking of the form $Y = X + \varepsilon$ where ε is a random vector with zero mean, covariance proportional to that of X, and whose probability distribution is a mixture of k normal distributions. The number k must exceed the dimension (number of variables) in the data X. The total covariance of ε is such that the $\mathrm{Cov}(Y) = (1 + d)\,\mathrm{Cov}(X)$ where $d, 0 < d < 1$, is pre-specified. The mean parameters of the component distributions are solved by a nonlinear optimization method. With the empirical data used by Roque, the bias in the individual component means have the effect of making re-identification more difficult in contrast to re-identification when the simple normal noise method of Kim is used.

In this paper, we provide a simpler computational approach using factorization of $\mathrm{Cov}(X)$. The advantage is that no nonlinear optimization solver needs to be applied. In fact the basic computational methods are a straightforward variant of the methods used in [10]. The appendix gives more details.

Note that since additive noise preserves means μ and produces a scalar inflation of the covariance matrix, we can rescale the masked data records y from the masked data set Y by

$$y' = \frac{1}{\sqrt{1+d}}y - \left(1 - \frac{1}{\sqrt{1+d}}\right)\mu \tag{1}$$

so that the scaled data set Y' has expected value mean μ and $\mathrm{Cov}\,(Y') = \mathrm{Cov}\,(X)$.

3.4   Re-identification

A record linkage process attempts to classify pairs in a product space $A \times B$ from two files A and B into M, the set of true links, and U, the set of true nonlinks. Fellegi and Sunter [6] considered ratios R of probabilities of the form

$$R = \frac{\Pr(\gamma \in \Gamma | M)}{\Pr(\gamma \in \Gamma | U)} \tag{2}$$

where γ is an arbitrary agreement pattern in a comparison space Γ. For instance, Γ might consist of eight patterns representing simple agreement or not on surname, first name, and age. Alternatively, each $\gamma \in \Gamma$ might additionally account for the relative frequency with which specific surnames, such as Scheuren or Winkler, occur or deal with different types of comparisons of quantitative data. The fields compared (surname, first name, age) are called matching variables. The numerator in (2) agrees with the probability given by equation (2.11) in [7]

The decision rule is given by:

1. If $R > T_\mu$, then designate pair as a link.
2. If $T_\lambda \leq R \leq T_\mu$, then designate pair as a possible link and hold for clerical review.
3. If $R < T_\lambda$, then designate pair as a nonlink.

Fellegi and Sunter [6] showed that this decision rule is optimal in the sense that for any pair of fixed bounds on R, the middle region is minimized over all decision rules on the same comparison space Γ. The cutoff thresholds, T_μ and T_λ, are determined by the error bounds. We call the ratio R or any monotonically increasing transformation of it (typically a logarithm) a matching weight or total agreement weight. Likely re-identifications, called matches, are given higher weights, and other pairs, called nonmatches, are given lower weights.

In practice, the numerator and denominator in (1) are not always easily estimated. The deviations of the estimated probabilities from the true probabilities can make applications of the decision rule suboptimal. Fellegi and Sunter [6] were the first to observe that

$$\Pr(\gamma \in \varGamma) = \Pr(\gamma \in \varGamma | M) \Pr(M) + \Pr(\gamma \in \varGamma | U) \Pr(U) \tag{3}$$

could be used in determining the numerator and denominator in (2) when the agreement pattern γ consists of simple agreements and disagreements of three variables and a conditional independence assumption is made. The left hand side is observed and the solution involves seven equations with seven unknowns. In general, we use the Expectation-Maximization (EM) algorithm [1] to estimate the probabilities on the right hand side of (3). To best separate the pairs into matches and nonmatches, our version of the EM algorithm for latent classes [16] determines the best set of matching parameters under certain model assumptions which are valid with the generated data and not seriously violated with the real data. In computing partial agreement probabilities for quantitative data, we make simple univariate adjustments to the matching weights such as are done in commercial record linkage software. When two quantitative items a and b do not agree exactly, we use a linear downward adjustment from the agreement matching weight to the disagreement weight according to a tolerance. For this analysis, we experimented with two methods of weight adjustment. For the raw data value a and the masked data value b, the d method adjustment is

$$w_{adj} = \max\left(\left\{w_{adj} - \frac{(w_{agr} - w_{dis})|a - b|}{t \max(|a|, 0.1)}\right\}, w_{dis}\right),$$

and the l method adjustment is

$$w_{adj} = \max\left(\left\{w_{adj} - \frac{(w_{agr} - w_{dis})|\log a - \log b|}{t \max(|\log a|, 0.1)}\right\}, w_{dis}\right),$$

where $w_{adj}, w_{agr}, w_{dis}$ are the adjusted weight, full agreement weight, and full disagreement weights, respectively and t is the proportional tolerance for the deviation $(0 \le t \le 1)$. The full agreement weights w_{agr} and disagreement weights w_{dis} are the natural logarithms of (2) that are obtained via the EM algorithm. The approximation will not generally yield accurate match probabilities but works well in the matching decision rules as we show later in this paper. Because we do not accurately account for the probability distribution with the generated multivariate normal data, our probabilities will not necessarily perform as well as the true probabilities used by Fuller when we consider single pairs. We note that the d method is basically a natural linear interpolation formula based on the distance between the raw and masked values. The l method was originally devised by Kim and Winkler [11] for multiplicative noise masking, but it has proven generally more effective than the d method for the additive noise case. This is probably because the most identifiable records are the ones containing large outlier values, and the l method tends to downweight less than the d method when both the input values are large.

To force 1-1 matching as an efficient global approach to matching the entire original data sets with the entire masked data sets, we apply an assignment algorithm due to [16]. Specifically, we use pairs $(i, j) \in I_0$ where I_0 is given by

$$I_0 = \min \left\{ \sum_{(i,j) \in I} w_{ij} \,\middle|\, I \subset J \right\},$$

where w_{ij} is the comparison weight for record pair (i, j), and J is the set of index sets I in which at most one column and at most one row are present. That is, if $(i, j) \in I$ and $(k, l) \in I$, then either $i \neq k$ or $j \neq l$. The algorithm of Winkler is similar to the classic algorithm of Burkard and Derigs (see *e.g.*, [16]) in that it uses Dijkstra's shortest augmenting path for many computations and has equivalent computational speed. It differs because it contains compression/decompression routines that can reduce storage requirements for the array of weights w_{ij} by a factor of 500 in some matching situations. When a few matching pairs in a set can be reasonably identified, many other pairs can be easily identified via the assignment algorithm. The assignment algorithm has the effect of drastically improving matching efficacy, particularly in re-identification experiments of the type given in this paper. For instance, if a moderate number of pairs associated with true re-identifications have probability greater than 0.5 when looked at in isolation, the assignment algorithm effectively sets their match probabilities to 1.0 because there are no other suitable records with which the truly matching record should be combined.

The proportion re-identified is the re-identification risk (PLD), computed using an updated version of the probabilistic re-identification software that has been used in [10], [14], and [5]. Domingo-Ferrer *et al.* [5] also used distance-based record linkage (DLD) for situations in which Euclidean distance is used. DLD can be considered a variant of nearest-neighbor matching. Because this DLD is highly correlated with record linkage [5], we only use PLD. Domingo-Ferrer *et al.* also used interval disclosure (ID) that we do not believe is appropriate because it is far too weak a disclosure-risk. See [5] for a definition of ID.

For the two data sets examined below, we counted the number of re-identified matches used to compute the re-identification risk somewhat differently. For the Domingo-Ferrer data, since the data set is small, we counted all of the correct matches in the set of linked pairs reported by the record linkage software. For the larger Kim-Winkler data set, a more realistic count of correctly re-identified matches is given by counting matches with agreement weights above a cutoff value T_μ, where the rest of the correct matches in the file are scattered sparsely among a large number of incorrect match pairs. In practice, these sporadic matches would not be detectable and their inclusion would produce an intolerably high false match rate.

3.5   Information-Loss and Scoring Metric

In [5] a number of formulas are suggested for measuring the "information loss" or amount which the masked data set has been statistically altered from the original raw data set. The idea is to compute some kind of penalty score to indicate how much the masked data set statistically differs from the original. The problem becomes one of identifying what one considers to be significant statistical properties and then to define a way to compute their difference for two data sets.

For original data set X and masked data set Z, both $n \times m$ arrays, one might want to consider a measure of the change in data, *i.e.* a measurement for $Z - X$. The original suggestion was for something like

$$\frac{1}{mn} \sum_{j=1}^{m} \sum_{i=1}^{n} \frac{|x_{ij} - z_{ij}|}{|x_{ij}|}$$

but this is undefined whenever we have an original data element $x_{ij} = 0$. One can replace the denominator by a specified constant when $x_{ij} = 0$, but then the value of this score can vary greatly with the choice of constant, especially when the data set has a lot of zero elements, as in the Kim-Winkler data. Furthermore, the size of this data perturbation score tends to be several orders of magnitude larger than the other information loss scores described below, so that it totally dominates all of the other scores when they are combined. Initially we tried to improve this situation by modifying the above formula to

$$IL1 = \frac{1}{mn} \sum_{j=1}^{m} \sum_{i=1}^{n} \frac{|x_{ij} - z_{ij}|}{0.5 \left(|x_{ij}| + |z_{ij}| \right)}$$

which helps with the score magnitude problem, since it is now bounded by $IL1 \leq 2$, but it only reduces but does not eliminate the possibility of dividing by zero. However, beyond these problems, by scaling the statistic by the individual data points, the resulting statistic is not very stable. Leaving aside zero data values, when the data values are very small, then small adjustments in the masked data produce large effects on the summary statistic. If we view our data set as independent samples from a common distribution, it is more stable to measure variations in the sample values by scaling them all by a value common to the variable. Thus if X and Z are independent random variables both with mean μ and variance σ^2, then the random variable $Y = X - Z$ has mean 0 and variance $2\sigma^2$. Hence a common scale for Z would be its standard deviation $\sqrt{2}\sigma$. In our case, we can estimate that standard deviation with the sample standard deviation S. This motivates the proposed modification for the data perturbation information loss statistic given by

$$IL1s = \frac{1}{mn} \sum_{j=1}^{m} \sum_{i=1}^{n} \frac{|x_{ij} - y_{ij}|}{\sqrt{2}S_j}.$$

This uses a common scale for all values of the same variable in the data set, the denominator is not zero unless the values of the variable are constant throughout the data set, and while the statistic does not have an *a priori* upper bound, the values in our empirical studies tend to be closer in magnitude to the other information loss statistics.

In summary we would suggest one of two approaches for a data perturbation score in the context of information loss measures for masked data sets. One approach would be to leave it out entirely. If the ideal of data masking is to try to preserve statistical properties of a data set while making individual records difficult to identify, when we are trying to assess the degree to which the statistical properties are preserved, perhaps we should not include a measure of how much individual records have been perturbed. The other approach, if one does want to include such a measure, then it would be better to use a more uniform and intrinsic scaling method, such as in $IL1s$.

The other information loss statistics that we compute are the same as some of those suggested by Domingo-Ferrer. To measure the variation in the sample means, we compute

$$IL2 = \frac{1}{m} \sum_{j=1}^{m} \frac{|\bar{x}_j - \bar{y}_j|}{|\bar{x}_j|}.$$

In theory, this score could also have the problem of zero or relatively small denominators, but since the sample means for our data sets are summary statistics for nonnegative whole real numbers, this did not seem to be a problem.

For variations in the sample covariance matrix, we compute

$$IL3 = \frac{2}{m(m+1)} \sum_{j=1}^{m} \sum_{k=1}^{j} \frac{\left| \text{Cov}(X)_{jk} - \text{Cov}(Y)_{jk} \right|}{\left| \text{Cov}(X)_{jk} \right|}$$

for variations in the sample variances, we compute

$$IL4 = \frac{1}{m} \sum_{j=1}^{m} \frac{\left| \text{Cov}(X)_{jj} - \text{Cov}(Y)_{jj} \right|}{\left| \text{Cov}(X)_{jj} \right|}$$

and for variations in the sample correlation matrix, we compute

$$IL5 = \frac{2}{m(m-1)} \sum_{j=1}^{m} \sum_{k=1}^{j-1} \left| \text{Cor}(X)_{jk} - \text{Cor}(Y)_{jk} \right|.$$

We wish to combine these information loss statistics into a summary information loss score. While it's not clear what sense it really makes to combine these numbers, and even if we do, its not clear what appropriate weighting we should give to them, in the absence of deeper insight, we just compute a straight average. However, we may choose which statistics we wish to include. As we have noted, the data perturbation measure $IL1$ is somewhat numerically problematic. Moreover, for the purposes of data masking, it is not clear if one cares how much individual data records are perturbed as long as the overall statistical structure of the data set is preserved. Thus one information loss penalty score can be computed by leaving out $IL1$ to get

$$s0 = \frac{IL2 + IL3 + IL4 + IL5}{4}.$$

On the other hand, one can leave it in to get

$$s1 = \frac{IL1 + IL2 + IL3 + IL4 + IL5}{5}.$$

Another objection to combining all theses scores is that $IL3, IL4$, and $IL5$ are redundant. With the covariance, variance, and correlation, if we know two of these things, then we know the third. Furthermore, the covariance score $IL3$ is to a lesser degree subject to the same kind of scaling instability as found with $IL1$, namely that the smallest values make the largest contributions to the score. In particular, we observe that the score tends to be dominated by those components corresponding to the smallest correlations. Thus as an alternative summary information loss statistic, we suggest using the rescaled data perturbation score and leaving out the covariance score to get

$$s2 = \frac{IL1s + IL2 + IL4 + IL5}{4}.$$

For comparison purposes, for the empirical results, we combine each of these information loss scores with the re-identification score $reid$, as discussed in Section 3.4 to obtain an overall data masking score. Specifically, the resulting scores are given by

$$Ascore = 100 \left(\frac{S0 + reid}{2} \right)$$

$$Dscore = 100 \left(\frac{S1 + reid}{2} \right)$$

$$Sscore = 100 \left(\frac{S2 + reid}{2} \right)$$

4   Results

4.1   Domingo Data Statistics

The Domingo data set consists of 1080 records from which we have masked 13 real variables. We can observe in Table 1 how the information loss scores increase with increasing noise level d. Rescaling the masked data has no mathematical effect on the mean and correlation. Since the rescaling somewhat contracts the data, there tends to be some decrease in the data perturbation scores. The effects of rescaling are most significant in the covariance and especially the variance scores.

Table 1. Domingo data information loss statistics

	IL1	IL1s	IL2	IL3	IL4	IL5	s0	s1	s2
rnkswp05	0.129	0.091	0.000	0.130	0.000	0.016	0.036	0.055	0.027
rnkswp10	0.219	0.155	0.000	0.195	0.000	0.036	0.058	0.090	0.048
rnkswp15	0.294	0.208	0.000	0.224	0.000	0.070	0.073	0.118	0.069
add01	0.194	0.137	0.001	0.036	0.014	0.002	0.013	0.038	0.039
add10	0.371	0.263	0.004	0.168	0.115	0.007	0.073	0.114	0.097
mixadd01	0.204	0.063	0.002	0.028	0.012	0.002	0.011	0.050	0.020
mixadd05	0.326	0.140	0.004	0.088	0.053	0.004	0.037	0.095	0.050
mixadd10	0.398	0.199	0.006	0.152	0.105	0.005	0.067	0.133	0.079
mixadd20	0.489	0.281	0.008	0.273	0.207	0.007	0.124	0.189	0.126
scalmixadd01	0.202	0.063	0.002	0.021	0.003	0.002	0.007	0.046	0.017
scalmixadd05	0.316	0.137	0.004	0.048	0.007	0.004	0.016	0.076	0.038
scalmixadd10	0.379	0.190	0.006	0.067	0.010	0.005	0.022	0.093	0.053
scalmixadd20	0.449	0.258	0.008	0.095	0.014	0.007	0.031	0.114	0.072

The matching software has two methods, the d method and the l method as discussed in Section 3.4 for measuring agreement between two real values. In either case, it interpolates between the agreement weight and the disagreement weight. In Table 2, we see that when the perturbations are small, the d method does a little better than the l method. However, when the perturbations get large, the l method is better able to see past moderate perturbations to large values.

The re-identification software produces a list of linked pairs in decreasing matching weight. For this small data set, the re-identification rate is computed as the total number of correctly linked pairs out of the total number of records in the data file. This is a rather optimistic re-identification score since most of the true matches are mixed among many false matches, and an analyst would probably have difficulty picking many of them out. In any event, we can see that for this data set with so few records and so many matching variables, a 1% noise level does not provide adequate masking, but the re-identification rate drops off rapidly with increasing noise level.

For an overall data masking score, we combine the information loss score with the re-identification score. Since we computed three data loss scores, we compute three overall scores in Table 3.

4.2 Kim-Winkler Data Statistics

The Kim-Winkler data consists of 59,315 records each containing 11 real variables for income data. In Table 4 we show the information loss statistics for our additive mixed noise masking for the whole data set. We note that the $IL1$ date perturbation statistic tends to higher than that for the Domingo data, possibly due to the large number of zero entries in the Kim-Winkler data, whereas the $IL1s$ data perturbation metric is about the same as for the Domingo data. In general the scaled data tends to get better results reducing the covariance measures $IL3, IL4$ than in the Domingo data case.

Table 2. Domingo data reidentification rates

	d metric	l metric
rnkswp05	0.8861	0.9620
rnkswp10	0.2694	0.7287
rnkswp15	0.0491	0.3444
add05	0.7972	0.7500
add10	0.2296	0.3167
mixadd01	0.7667	0.7176
mixadd05	0.1482	0.3556
mixadd10	0.0574	0.2194
mixadd20	0.0139	0.1009
scalmixadd01	0.7704	0.7370
scalmixadd05	0.1602	0.3537
scalmixadd10	0.0648	0.2417
scalmixadd20	0.0269	0.1241

Table 3. Domingo data scoring metrics

	d Metric			l Metric		
	Ascore	Dscore	Sscore	Ascore	Dscore	Sscore
rnkswp05	46.11	47.06	46.66	49.90	50.85	49.45
rnkswp10	16.37	17.97	15.87	39.34	40.94	38.84
rnkswp15	6.11	8.36	5.91	20.87	23.12	30.67
add01	40.51	41.76	41.81	38.15	39.40	39.45
add10	15.13	17.18	16.33	19.49	21.54	20.69
mixadd01	38.88	40.81	39.31	36.42	38.36	36.86
mixadd05	9.27	12.16	9.93	19.64	22.53	20.30
mixadd10	6.22	9.53	6.80	14.32	17.63	14.90
mixadd20	6.88	10.13	6.98	11.23	14.48	11.33
scalmixadd01	38.95	40.90	39.46	37.21	39.16	37.72
scalmixadd05	8.94	11.94	10.06	18.47	21.47	19.59
scalmixadd10	4.43	8.00	5.97	13.19	16.75	14.73
scalmixadd20	2.89	7.06	4.94	7.75	11.92	9.80

Table 4. Kim-Winkler data information loss statistics, 11 variables

	IL1	IL1s	IL2	IL3	IL4	IL5	s0	s1	s2
mixadd01	1.165	0.060	0.002	0.014	0.010	0.000	0.007	0.238	0.018
mixadd05	1.308	0.135	0.005	0.056	0.051	0.001	0.028	0.284	0.048
mixadd10	1.381	0.191	0.007	0.106	0.101	0.001	0.054	0.319	0.075
mixadd20	1.457	0.270	0.010	0.201	0.201	0.002	0.104	0.374	0.121
scalmixadd01	1.163	0.060	0.002	0.008	0.001	0.000	0.003	0.235	0.016
scalmixadd05	1.302	0.132	0.005	0.018	0.002	0.001	0.006	0.265	0.035
scalmixadd10	1.370	0.183	0.007	0.025	0.002	0.001	0.009	0.281	0.048
scalmixadd20	1.438	0.249	0.010	0.033	0.003	0.002	0.012	0.297	0.066

Table 5. Kim-Winkler data information loss statistics, 8 variables

	IL1	IL1s	IL2	IL3	IL4	IL5	s0	s1	s2
rnkswp05	0.174	0.123	0.000	0.525	0.000	0.197	0.180	0.179	0.080
rnkswp10	0.280	0.198	0.000	0.609	0.000	0.211	0.205	0.220	0.102
rnkswp15	0.362	0.256	0.000	0.605	0.000	0.214	0.205	0.236	0.118
add01	1.271	0.897	0.006	0.018	0.009	0.019	0.013	0.331	0.235
add01_sw	1.286	0.909	0.006	0.018	0.009	0.009	0.013	0.335	0.232
mixadd01	1.304	0.061	0.003	0.012	0.010	0.000	0.006	0.266	0.019
mixadd05	1.443	0.137	0.006	0.052	0.051	0.001	0.027	0.311	0.049
mixadd10	1.512	0.194	0.008	0.101	0.101	0.001	0.053	0.345	0.076
mixadd20	1.582	0.274	0.012	0.199	0.201	0.002	0.103	0.397	0.122
scalmixadd01	1.302	0.061	0.003	0.005	0.001	0.000	0.002	0.262	0.016
scalmixadd05	1.437	0.134	0.006	0.011	0.002	0.001	0.005	0.291	0.037
scalmixadd10	1.503	0.185	0.008	0.015	0.002	0.001	0.007	0.306	0.049
scalmixadd20	1.567	0.252	0.012	0.020	0.003	0.002	0.009	0.321	0.067

For our re-identification, we only used eight of the income variables, so we computed the information loss scores based on just these eight variables. In Table 5, they show a generally slight increase over the eleven variable scores. For the re-identification scores, we computed the proportion of correctly linked pairs out of the total number of records, as in the case of the Domingo data. In this case, we only compute the results using the l interpolation metric in Table 6, since it is much more effective on this data. However, reporting the total number of correct matches in the full link file is probably even more misleadingly optimistic than in the Domingo data case. For the Domingo data, the true matches tend to be distributed throughout the link file. As the noise level of the masking increases, this distribution becomes more sparse and random. In the case of this data set, there are twenty or so records that are extreme outliers with one or more income categories much higher than the values for the mass of the records. Many of these records fail to be successfully masked from the re-identification through most noise levels, especially using the l metric. Thus there are always several clearly true matches at the top of the match-weight sorted link file. However, as the matching weights decrease, the proportion of true matches rapidly drops off as we include more and more false links in with a decreasing number of true matches. Thus is seems reasonable to cut off the count of true matches at some point, since beyond this point, any true matches will only appear sporadically among the preponderance of false matches and are unlikely to be discerned by the analyst. Here we choose a rather low cutoff point of 20%. This means that at this point, the number of linked pairs at this matching weight or higher contain 20% true matches and 80% false links. Below this point, the true matches become much rarer. In Table 7 are the overall data masking scores for the Kim-Winkler data. The data masking tends to be more effective here, especially at lower additive noise levels. Again inclusion of the $IL1$ data perturbation score tends to dominate and obscure the rest of the results.

Table 6. Kim-Winkler data reidentification rates, l metric

	Total File Matches	20% Zone
rnkswp05	0.8032	0.8032
rnkswp10	0.6072	0.6072
rnkswp15	0.4855	0.4855
add01	0.0590	0.0420
add01_sw	0.0010	0.0000
mixadd01	0.0841	0.0960
mixadd05	0.0346	0.0027
mixadd10	0.0240	0.0018
mixadd20	0.0149	0.0011
scalmixadd01	0.0844	0.0098
scalmixadd05	0.0355	0.0031
scalmixadd10	0.0249	0.0022
scalmixadd20	0.0174	0.0016

Table 7. Kim-Winker data scoring metrics

	Full File Matches			20% Zone Matches		
	Ascore	Dscore	Sscore	Ascore	Dscore	Sscore
rnkswp05	98.35	98.21	88.32	98.35	98.21	88.32
rnkswp10	81.23	82.72	70.94	81.23	82.72	70.94
rnkswp15	69.03	72.17	60.81	69.03	72.17	60.81
add01	3.60	19.50	14.70	2.75	18.75	13.85
add01_sw	0.70	16.80	11.65	0.65	16.75	11.60
mixadd01	4.52	17.49	5.14	0.80	13.77	1.41
mixadd05	3.10	17.26	4.16	1.51	15.66	2.57
mixadd10	3.85	18.44	5.00	2.74	17.33	3.84
mixadd20	5.92	19.45	6.78	5.23	18.76	6.09
scalmixadd01	4.33	17.33	5.03	0.60	13.60	1.30
scalmixadd05	2.02	16.35	3.62	0.40	15.45	2.00
scalmixadd10	1.58	16.54	3.71	0.45	15.41	2.58
scalmixadd20	1.33	16.91	4.22	0.49	16.12	3.43

Subpopulation Information Loss Statistics. The additive noise procedures are supposed to preserve means and covariances on arbitrary subpopulations, at least when these statistics are properly corrected, according to the method of Kim [9]. In Tables 8 and 9 we compute the information loss scores for two subpopulations. We see that even for the corrected means and covariances, there are still generally somewhat higher scores than for the full data set. We especially note that the scaled data sets fail to recover the original data covariance and variance values as well.

In Table 9 we see slightly better information loss scores for a slightly larger subpopulation.

Table 8. S4 return type information loss, 8 variables, 5885 records

	IL1	IL1s	IL2	IL3	IL4	IL5	s0	s1	s2
rnkswp05	0.114	0.081	1.407	39.020	158.950	0.123	48.875	39.923	40.141
rnkswp10	0.199	0.141	0.397	1.682	3.980	0.179	1.560	1.287	1.174
rnkswp15	0.277	0.196	0.151	0.918	0.799	0.174	0.511	0.464	0.330
add01	1.364	0.964	0.027	0.343	0.240	0.017	0.156	0.398	0.312
add_sw01	1.364	0.964	0.027	0.343	0.240	0.017	0.156	0.398	0.312
mixadd01	1.399	0.246	0.057	0.101	0.008	0.006	0.043	0.314	0.079
mixadd05	1.559	0.550	0.128	0.232	0.024	0.014	0.099	0.391	0.179
mixadd10	1.631	0.777	0.181	0.342	0.040	0.021	0.146	0.443	0.255
mixadd20	1.697	1.099	0.256	0.539	0.069	0.033	0.224	0.519	0.364
scalmixadd01	1.397	0.245	0.057	0.130	0.008	0.006	0.050	0.320	0.079
scalmixadd05	1.552	0.536	0.128	0.298	0.024	0.014	0.116	0.403	0.175
scalmixadd10	1.621	0.741	0.181	0.440	0.040	0.021	0.170	0.460	0.246
scalmixadd20	1.681	1.004	0.256	0.693	0.069	0.033	0.263	0.547	0.341

Table 9. Schedule C subset information loss, 8 variables, 7819 records

	IL1	IL1s	IL2	IL3	IL4	IL5	s0	s1	s2
rnkswp05	0.207	0.146	0.082	0.711	0.523	0.250	0.391	0.355	0.250
rnkswp10	0.322	0.228	0.125	0.796	0.573	0.261	0.438	0.415	0.297
rnkswp15	0.410	0.290	0.126	0.765	0.473	0.266	0.408	0.408	0.299
add01	1.221	0.863	0.013	0.021	0.017	0.003	0.013	0.255	0.224
add01_sw	1.250	0.884	0.057	0.431	0.296	0.084	0.217	0.424	0.330
mixadd01	1.410	0.220	0.012	0.065	0.009	0.005	0.023	0.288	0.062
mixadd05	1.560	0.492	0.027	0.175	0.026	0.012	0.060	0.360	0.139
mixadd10	1.627	0.696	0.038	0.299	0.043	0.019	0.100	0.405	0.199
mixadd20	1.688	0.984	0.054	0.531	0.075	0.030	0.172	0.476	0.286
scalmixadd01	1.408	0.219	0.012	0.084	0.009	0.005	0.028	0.304	0.061
scalmixadd05	1.553	0.480	0.027	0.225	0.026	0.012	0.073	0.369	0.136
scalmixadd10	1.617	0.664	0.038	0.385	0.043	0.019	0.121	0.420	0.191
scalmixadd20	1.674	0.899	0.054	0.682	0.075	0.030	0.210	0.503	0.265

5   Discussion

For the research community, there are two general difficulties with comparing masking methods. The first is that the suitable test files are needed. The test files should have variables in which the distributions are representative of actual databases. Some of the test files should have quite skewed distributions. Others should have a large number of zeros for several of the variables The second is that the information-loss metrics should be reasonably robust across different types of databases. We observed that some of the metrics that we have used in this paper are sensitive to the skewness of distributions and the proportions of zeros associated with a variable.

Much of prior research (e.g., [8], [7], [10]) dealt with situations where only a few specific analyses were demonstrated to be approximately reproduced with a masked data file. In most of the situations, special software was needed to do many of the analyzes, particularly on subdomains. If masked files are required to reproduced more than one or two sets of analyses from original, unmasked data, then we suspect the special methods and software will be typically needed.

6   Concluding Remarks

This paper provides a comparison of rank swapping with various methods of additive noise. In the comparison of [5], rank swapping provided the best trade-off between information-loss and disclosure risk with measures used in the earlier work. With the same data and the same metrics, rank swapping provides better results than the types of mixtures of additive noise that we provide in this paper. With other, much larger data [10] that represents actual public-use situations, scaled mixtures of additive noise perform best with the same scoring metrics. For the scoring methods used here, using scaled masked data produces improved scores since the information loss scores are improved by better covariance agreement while the re-identification risk is only slightly worse. This suggests that additional scoring metrics and more applications to different data situations are needed. The scoring metrics, particularly the components of information loss, need to be better connected to additional analyzes and specific characteristics of data.

References

1. Dempster, A.P., Laird, N.M., and Rubin, D.B.: Maximum Likelihood from Incomplete Data via the EM Algorithm, *Journal of the Royal Statistical Society*, **B, 39** (1977) 1–38.
2. Dalenius, T. and Reiss, S.P.: Data-swapping: A Technique for Disclosure Control of Microdata, *Journal of Statistical Planning and Inference*, **6** (1982) 73–85.
3. De Waal, A.G. and Willenborg, L.C.R.J.: A View on Statistical Disclosure Control of Microdata, *Survey Methodology*, **22**, (1996) 95–103.
4. De Waal, A.G. and Willenborg, L.C.R.J.: Optimal Local Suppression in Microdata, *Journal of Official Statistics*, **14**, (1998) 421–435.
5. Domingo-Ferrer, J., Mateo-Sanz, J., and Torra, V.: Comparing SDC Methods for Microdata on the Basis of Information Loss and Disclosure Risk, *Proceedings of ETK-NTTS* 2001, (2001) to appear.
6. Fellegi, I.P. and Sunter, A.B.: A Theory for Record Linkage, *Journal of the American Statistical Association*, **64**, (1969) 1183–1210.
7. Fuller, W.A.: Masking Procedures for Microdata Disclosure Limitation, *Journal of Official Statistics*, **9**, (1993) 383–406.
8. Kim, J.J.: A Method for Limiting Disclosure in Microdata Based on Random Noise and Transformation, *American Statistical Association, Proceedings of the Section on Survey Research Methods*, (1986) 303–308.
9. Kim, J.J.: Subdomain Estimation for the Masked Data, *American Statistical Association, Proceedings of the Section on Survey Research Methods*, (1990) 456–461.
10. Kim, J.J. and Winkler, W.E.: Masking Microdata Files, *American Statistical Association, Proceedings of the Section on Survey Research Methods*, (1995) 114–119.

11. Kim, J.J. and Winkler, W.E.: Multiplicative Noise for Masking Continuous Data, *American Statistical Association Proceedings of Secure Survey Research Methods*, (to appear).
12. Lambert, D.: Measures of Disclosure Risk and Harm, *Journal of Official Statistics*, **9**, (1993) 313–331.
13. Moore, R.: Controlled Data Swapping Techniques for Masking Public Use Microdata, *U.S. Bureau of the Census, Statistical Research Division Report 96/04 (1996)*.
14. Roque, G.M.: *Masking Microdata Files with Mixtures of Multivariate Normal Distributions*, Unpublished Ph.D. dissertation, Department of Statistics, University of California–Riverside (2000).
15. Tendick, P. and Matloff, N.: A Modified Random Perturbation Method for Database Security, *ACM Transactions on Database Systems*, **19**, (1994) 47–63.
16. Winkler, W.E.: Advanced Methods for Record Linkage, *American Statistical Association, Proceedings of the Section on Survey Research Methods*, (1994) 467–472.
17. Winkler, W.E.: Matching and Record Linkage, in B.G. Cox (ed.) *Business Survey Methods*, New York: J. Wiley, (1995) 355–384.
18. Winkler, W.E.: Re-identification Methods for Evaluating the Confidentiality of Analytically Valid Microdata, *Research in Official Statistics*, **1**, (1998) 87–104

Appendix: Additive Mixture Noise Methodology

As in the case of normal additive noise, to the given raw data set X, an $n \times m$ array, we wish to produce a masked data set Z by adding a masking noise array Y

$$Z = X + dY$$

where the records of Y are independent samples of a distribution with zero mean and $\text{Cov}(Y) = \text{Cov}(X) = \Sigma$. Typically for additive noise we chose a normal distribution $N(0, \Sigma)$; for mixture noise we may choose a normal mixture distribution $\sum_{k=1}^{K} \omega_k N(\theta_k, \Sigma_k)$. For a probability distribution, the weights are constrained so that $\sum_{k=1}^{K} \omega_k = 1, \omega_k > 0$. To obtain zero mean, we must have $\sum_{k=1}^{K} \omega_k \theta_k = 0$. When we choose $\Sigma_k = \sigma_k \Sigma$, for $K > m$, in general to obtain total covariance Σ, the component means θ_k must further satisfy a (underdetermined) system of quadratic equations that can be solved numerically, as addressed in [14]. However, it is computationally simpler to produce colored noise from white noise. That is, if $\Sigma^{\frac{1}{2}}$ is a square root of the positive definite symmetric matrix Σ,

$$\Sigma = \Sigma^{\frac{1}{2}} \left(\Sigma^{\frac{1}{2}} \right)^T$$

and w is an uncorrelated random vector with mean 0 and identity covariance I, then the random vector y,

$$y = \Sigma^{\frac{1}{2}} w$$

has mean 0 and covariance Σ. To produce w, we need m independent components w_j of mean 0 and variance 1. For standard normal additive noise, we can choose each w_j to be distributed as $w_j \sim N(0,1)$; for mixture distribution noise, we may choose each w_j to be distributed as

$$w_j \sim \sum_{k=1}^{K} \omega_k N\left(\theta_k, \sigma^2\right)$$

for some choice weights and common variance $\sigma^2 < 1$. Such a mixture distribution has mean $\sum_{k=1}^{K} \omega_k \theta_k$ and variance $\sigma^2 + \sum_{k=1}^{K} \omega_k \theta_k^2$. To obtain component means θ_k so that the mixture distribution has mean 0 and variance 1, we can start out with arbitrary numbers $\psi_1, \psi_2, \ldots, \psi_{K-1}$ and let

$$\psi_k = -\frac{1}{\omega_K} \sum_{k=1}^{K-1} \omega_k \psi_k$$

and compute

$$S = \sum_{k=1}^{K} \omega_k \psi_k^2$$

and let

$$\theta_k = \sqrt{\frac{1 - \sigma^2}{S}} \psi_k$$

The simplest case occurs when we choose the weights ω_k to be equal and the smallest number of components $K = 2$. In this case we have $\theta_1 = \sqrt{1 - \sigma^2}$ and $\theta_2 = -\sqrt{1 - \sigma^2}$. The mixture distribution

$$\frac{1}{2} N\left(\sqrt{1 - \sigma^2}, \sigma^2\right) + \frac{1}{2} N\left(-\sqrt{1 - \sigma^2}, \sigma^2\right)$$

differs most from the standard normal distribution when we choose a value of σ^2 near 0, where we get a bimodal distribution with modes near ± 1. A weakness of using standard normal additive noise for masking is that most samples from the distribution tend to be near zero and hence produce small perturbation to the data. In this simplest mixture model, samples from the distribution tend to be near either 1 or -1, and thus should produce a more substantial data perturbation. For the data masking for these empirical studies, we used a value of $\sigma^2 = 0.025$.

Using this uncorrelated, zero mean mixture distribution, we generated a data set

$$W = \begin{pmatrix} w_1^T \\ w_2^T \\ \vdots \\ w_n^T \end{pmatrix}$$

where each vector w_i is drawn from the above zero mean, identity covariance mixture distribution. Using a square root $\Sigma^{\frac{1}{2}}$ of the sample covariance matrix of X, we compute a colored noise data set

$$Y = \begin{pmatrix} \left(\Sigma^{\frac{1}{2}} w_1 \right)^T \\ \left(\Sigma^{\frac{1}{2}} w_2 \right)^T \\ \vdots \\ \left(\Sigma^{\frac{1}{2}} w_n \right)^T \end{pmatrix}$$

and for different noise proportion parameters d, we compute a masked data set

$$Z = X + dY.$$

Since the resulting data set Z theoretically has covariance $(1 + d)\, \Sigma$, for each d we also compute a scaled masked data set Z_s related by

$$z_s = \frac{1}{\sqrt{1+d}} z + \left(1 - \frac{1}{\sqrt{1+d}} \right) \mu$$

where μ is the (sample) mean vector of the masked data.

LHS-Based Hybrid Microdata vs Rank Swapping and Microaggregation for Numeric Microdata Protection*

Ramesh A. Dandekar[1], Josep Domingo-Ferrer[2], and Francesc Sebé[2]

[1] Energy Information Administration, U. S. Department of Energy
1000 Independence Ave, Washington DC 20585, USA
ramesh.dandekar@eia.doe.gov
[2] Universitat Rovira i Virgili,Department of Computer Science and Mathematics
Av. Països Catalans 26, E-43007 Tarragona, Catalonia, Spain
{jdomingo,fsebe}@etse.urv.es

Abstract. In previous work by Domingo-Ferrer *et al.*, rank swapping and multivariate microaggregation has been identified as well-performing masking methods for microdata protection. Recently, Dandekar et al. proposed using synthetic microdata, as an option, in place of original data by using Latin hypercube sampling (LHS) technique. The LHS method focuses on mimicking univariate as well as multivariate statistical characteristics of original data. The LHS-based synthetic data does not allow one to one comparison with original data. This prevents estimating the overall information loss by using current measures. In this paper we utilize unique features of LHS method to create hybrid data sets and evaluate their performance relative to rank swapping and multivariate microaggregation using generalized information loss and disclosure risk measures.

Keywords: Microdata masking, synthetic microdata generation, rank swapping, microaggregation.

1   Introduction

Statistical Disclosure Control (SDC) methods are used in official statistics to ensure confidentiality in statistical databases being released for public use[12]. If the database contains microdata (*i.e.* individual respondent records), confidentiality means trying to avoid disclosing the identity of the individual respondent associated with a published record. At the same time, SDC should preserve the informational content as much as possible. SDC methods are somewhere between encryption of the original data set (no disclosure risk but no informational content released) and straightforward release of the original dataset (no confidentiality but maximal informational content released). SDC methods for microdata are also known as masking methods.

---

* The second and third authors are partly supported by the European Commission under project IST-2000-25069 "CASC".

J. Domingo-Ferrer (Ed.): Inference Control in Statistical Databases, LNCS 2316, pp. 153–162, 2002.
© Springer-Verlag Berlin Heidelberg 2002

In [4,5], a comparison of masking methods for microdata was conducted. For numerical microdata, two masking methods were determined to be well-performing in that they achieve a good tradeoff between low disclosure risk and information loss. The first method is rank swapping [11] and the second method is multivariate microaggregation [6]. Both methods will be termed "natural" because they obtain a number of masked data records through transformation of the same number of original data records.

Dandekar *et al.* have proposed synthetic microdata generation as an option to natural masking methods [2]. Specifically, Latin hypercube sampling (LHS) has been used to obtain a synthetic masked data set with statistical properties similar to the original data set; masked data being synthetic, the number of masked data records does not need to be the same as the number of original data records.

We generate hybrid data by combining synthetic data with original sampled data. We then compare the performance of natural and hybrid masking by using several information loss and disclosure risk metrics. Section 2 recalls the essentials of rank swapping and multivariate microaggregation. Section 3 summarizes LHS synthetic data generation and describes several hybrid approaches to obtain masked data as a mixture of synthetic and original data. Section 4 discusses the metrics used to compare the three methods. Computational results are reported in Section 5. Finally, Section 6 is a conclusion.

2   Natural Masking Methods

We will summarize here the principles of rank swapping and multivariate microaggregation.

Rank swapping was originally described for ordinal variables, but it can actually be used for any numerical variable [11]. First, values of variable V_i are ranked in ascending order; then each ranked value of V_i is swapped with another ranked value randomly chosen within a restricted range (*e.g.* the rank of two swapped values cannot differ by more than $p\%$ of the total number of records). In [4,5], values p from 1 to 20 were considered in experimentation.

The basic idea of microaggregation is to cluster records into small aggregates or groups of size at least k [3,6]. Rather than publishing a variable for a given individual, the average of the values of the variable over the group to which the individual belongs is published. Variants of microaggregation include: individual ranking (MicIRk); microaggregation on projected data using z-scores projection (MicZk) and principal components projection (MicPCPk); microaggregation on unprojected multivariate data considering two variables at a time (Mic2mul$k$), three variables at a time (Mic3mul$k$), four variables at a time (Mic4mul$k$)or all variables at a time (Micmul$k$). Values of k between 3 and 18 have been considered. According to the experimental work [4,5], the best microaggregation variant is microaggregation on unprojected multivariate data considering three or four variables at a time.

3   Synthetic and Hybrid Masking Methods

Generation of synthetic data which preserves some of the characteristics of original data has been proposed as an alternative to natural masking methods. Multiple imputation [9] is one sophisticated approach to synthetic data generation which requires specific software not usually available. LHS synthetic data generation discussed below is a simpler approach which also yields satisfactory results.

3.1   LHS Synthetic Masking

Dandekar *et al.* [2] use Latin hypercube sampling (LHS) developed by [10] along with the rank correlation refinement of [7] to generate a synthetic data set which reproduces both the univariate and the multivariate structure of the original data set. The basic LHS algorithm generates a synthetic data set for a group of uncorrelated variables in which the univariate characteristics of the original data are reproduced almost exactly. In case the variables are not uncorrelated, the restricted pairing algorithm of [7], which makes use of LHS, is designed to produce a synthetic data set that reproduces the rank correlation structure of the original data[1].

3.2   Hybrid Masking Methods

While pure synthetic data may reproduce univariate and multivariate characteristics of the original data, one-to-one comparison of original and synthetic records poses problems. In other words, given an original record or a subset of original records, it is possible that there is no similar synthetic record or no similar subset of synthetic records. To minimize such a possibility, Dandekar *et al.* [2] recommends using the LHS procedure at subpopulation levels to the extent possible.

Hybrid masking consists of computing masked data as a combination of original and synthetic data. Such a combination allows better control over individual characteristics of masked records. Additive as well as multiplicative combinations could be used. For hybrid masking to be feasible, a rule must be used to pair one original data record with one synthetic data record. A sensible option is to go through all original data records, and pair each original data record with the nearest synthetic record according to some distance.

Example 1 (Euclidean Record Pairing). Assume an original data set consisting of n records, and a synthetic data set consisting of m records. Assume further that both data sets refer to the same d *numerical variables*. Then the d-dimensional Euclidean distance can be used for pairing in the following way:

1. First, variables in both data sets are standardized (subtract to the values of each variable their average value and divide them by their standard deviation).

2. Pair each record in the original standardized data set with the nearest record in the synthetic standardized data set, where "nearest" means at the smallest d-dimensional Euclidean distance.

Once the pairing of original and synthetic records has been done, we need a model to mix variables in paired records in order to get a hybrid masked data set. For numerical variables, at least two different hybrid models are conceivable for combining a variable in the original data set with the corresponding variable in the synthetic data set:

Definition 1 (Additive Hybrid Model). *Let X be a variable in the original data set. Let X_s be the variable corresponding to X in the synthetic data set. Let α be a real number in $[0,1]$. Then the additive hybrid masked version X_{ah} can be obtained from X and X_s as*

$$X_{ah} = \alpha X + (1 - \alpha)X_s \tag{1}$$

Definition 2 (Multiplicative Hybrid Model). *Let X be a variable in the original data set. Let X_s be the variable corresponding to X in the synthetic data set. Let α be a real number in $[0,1]$. Then the multiplicative hybrid masked version X_{mh} can be obtained from X and X_s as*

$$X_{mh} = X^\alpha \cdot X_s^{(1-\alpha)} \tag{2}$$

Note that the above pairing strategy yields as many pairs as there are original records. This implies that mixing pairs using hybrid models (1) and (2) will result in a masked data set with the same number of records as the original data set. Such a constraint is a flexibility loss in comparison to pure synthetic data sets, whose cardinality can be made independent from the cardinality of the original data sets. A way to remedy the above rigidity is to use *resampled* original data instead of original data. Assume that the original data set consists of n records and that an n'-record masked data set is desired. Then we can obtain the n' masked records by using the following algorithm:

Algorithm 1.   *1. Resample with replacement the n-record original data set to obtain an n'-record resampled data set.*
2. *For each of the n' resampled records, pair it with the nearest record in the synthetic data set, where nearest means at smallest d-dimensional Euclidean distance.*
3. *Within each record pair, mix corresponding variables using one of the models (1) or (2) above.*

4   Metrics for Method Comparison

In [5] a metric was proposed for method comparison. That metric needs some adaptation to deal with the case where the number of records in the original and the masked data sets are not the same. We first recall the [5] and then discuss some adaptations.

4.1 Same Number of Original and Masked Records

Let X and X' be the original and the masked data sets. Let V and V' be the covariance matrices of X and X', respectively; similarly, let R and R' be the correlation matrices. Table 1 summarizes the information loss measures proposed. In this table, d is the number of variables, n the number of records, and components of matrices are represented by the corresponding lowercase letters (e.g. x_{ij} is a component of matrix X). Regarding $X - X'$ measures, it also makes sense to compute those on the averages of variables rather than on all data (see the $\bar{X} - \bar{X}'$ row in Table 1). Similarly, for $V - V'$ measures, it is also sensible to compare only the variances of variables, i.e. to compare the diagonals of the covariance matrices rather than the whole matrices (see the $S - S'$ row in Table 1).

Table 1. Information loss measures

	Mean square error	Mean abs. error	Mean variation
$X - X'$	$\dfrac{\sum_{j=1}^{d}\sum_{i=1}^{n}(x_{ij}-x'_{ij})^2}{nd}$	$\dfrac{\sum_{j=1}^{d}\sum_{i=1}^{n}|x_{ij}-x'_{ij}|}{nd}$	$\dfrac{\sum_{j=1}^{d}\sum_{i=1}^{n}\frac{|x_{ij}-x'_{ij}|}{|x_{ij}|}}{nd}$
$\bar{X} - \bar{X}'$	$\dfrac{\sum_{j=1}^{d}(\bar{x}_{j}-\bar{x}'_{j})^2}{d}$	$\dfrac{\sum_{j=1}^{d}|\bar{x}_{j}-\bar{x}'_{j}|}{d}$	$\dfrac{\sum_{j=1}^{d}\frac{|\bar{x}_{j}-\bar{x}'_{j}|}{|\bar{x}_{j}|}}{d}$
$V - V'$	$\dfrac{\sum_{j=1}^{d}\sum_{1\le i\le j}(v_{ij}-v'_{ij})^2}{\frac{d(d+1)}{2}}$	$\dfrac{\sum_{j=1}^{d}\sum_{1\le i\le j}|v_{ij}-v'_{ij}|}{\frac{d(d+1)}{2}}$	$\dfrac{\sum_{j=1}^{d}\sum_{1\le i\le j}\frac{|v_{ij}-v'_{ij}|}{|v_{ij}|}}{\frac{d(d+1)}{2}}$
$S - S'$	$\dfrac{\sum_{j=1}^{d}(v_{jj}-v'_{jj})^2}{d}$	$\dfrac{\sum_{j=1}^{d}|v_{jj}-v'_{jj}|}{d}$	$\dfrac{\sum_{j=1}^{d}\frac{|v_{jj}-v'_{jj}|}{|v_{jj}|}}{d}$
$R - R'$	$\dfrac{\sum_{j=1}^{d}\sum_{1\le i<j}(r_{ij}-r'_{ij})^2}{\frac{d(d-1)}{2}}$	$\dfrac{\sum_{j=1}^{d}\sum_{1\le i<j}|r_{ij}-r'_{ij}|}{\frac{d(d-1)}{2}}$	$\dfrac{\sum_{j=1}^{d}\sum_{1\le i<j}\frac{|r_{ij}-r'_{ij}|}{|r_{ij}|}}{\frac{d(d-1)}{2}}$

Disclosure risk can be measured using record linkage. Two record linkage methods were combined in [5]:

Distance-based record linkage. Let the original and masked data sets consist both of d variables (it is assumed that both data sets contain the same variables). We define that a record in the masked data set corresponds to the nearest record in the original data set, where "nearest" means at shortest d-dimensional Euclidean distance. Assume further that the intruder can only access i key variables of the original data set and tries to link original and masked record based on these i variables. Linkage then proceeds by computing i-dimensional distances between records in the original and the masked data sets (distances are computed using only the i key variables) The variables used are standardized to avoid scaling problems. A record in the masked data set is labeled as "correctly linked" when the nearest record using i-dimensional distance is the corresponding one (i.e. the nearest record using d-dimensional distance).

Probabilistic record linkage. Defined in [8], uses a matching algorithm to pair records in the masked and original data sets. The matching algorithm is based on the linear sum assignment model. The definition of "correctly linked" records is the same as in distance-based record linkage. This method is attractive because it only requires the user to provide two probabilities as input: one is an upper bound of the probability of a false match and the other an upper bound of the probability of a false non-match. Unlike distance-based record linkage, probabilistic record linkage does not require rescaling variables nor makes any assumption on their relative weight (by default, distance-based record linkage assumes that all variables have the same weight). Furthermore, probabilistic record linkage can be used on both numerical and categorical data.

In [5], a score was constructed to rate methods which combined some of the above information loss and disclosure risk measures. The components of the score were as follows:

IL Information Loss: 100 times the average of the mean variation of $X - X'$ (called IL_1), the mean variation of $\bar{X} - \bar{X}'$ (called IL_2), the mean variation of $V - V'$ (called IL_3), the mean variation of $S - S'$ (called IL_4) and the mean absolute error of $R - R'$ (called IL_5).
DLD Average of DLD-1, $\cdots$, DLD-7. DLD-i is the percent of records correctly linked using distance-based record linkage with Euclidean distance when the intruder knows i key variables of the original file.
PLD Same as DLD, but for probabilistic record linkage.
ID Average percent of original values falling in the intervals around their corresponding masked values. The average is over interval widths $p = 1\%$ to $p = 10\%$.
Overall score

$$Score = 0.5 \cdot IL + 0.125 \cdot DLD + 0.125 \cdot PLD + 0.25 \cdot ID \qquad (3)$$

The lower $Score$, the better is a method.

4.2   Different Number of Original and Masked Records

Computation of the IL_1 component in the score described in Subsection 4.1 implicitly assumes that there exists a one-to-one mapping between original and masked records. For natural masking, the number of masked and original records is the same and it is (in theory) possible to track from which original record a masked record originates (for example using metadata such as record identifier); so a one-to-one mapping can be established.

When the number of masked records is not the same as the number of original records (as it can happen with synthetic or hybrid data), then there is no one-to-one mapping any more. In this case, a new way to compute IL_1 must be defined. A natural way is to map each published masked record to the nearest original record, using the d-dimensional Euclidean distance between records (where d is the number of variables in the data sets). Then we compute a new IL'_1 as the sum of differences between masked records and the original records to which they are mapped. Replacing IL_1 by IL'_1 leads to a modified information loss measure IL'.

Also, the lack of a one-to-one mapping between original and masked records forces a redefinition of disclosure risk measures DLD and PLD. If the masked and the original data sets have d variables, we will now say that a masked record is correctly linked to an original record if they are at the shortest possible d-dimensional Euclidean distance. Finally, ID can be redefined so that "corresponding values" mean values in records at shortest d-dimensional Euclidean distance. Call DLD', PLD' and ID' the resulting redefined disclosure risk measures.

Call $Score'$ the new score arising from replacing IL and DLD with IL' and DLD' in Equation 3 as well as dropping PLD for computational reasons:

$$Score' = 0.5 \cdot IL' + 0.25 \cdot DLD' + 0.25 \cdot ID' \tag{4}$$

5 Computational Results

Two original data sets have been tried:

Data set 1. This microdata set was constructed using the Data Extraction System (DES) of the U.S. Census Bureau[1]. $d = 13$ continuous variables were chosen and 1080 records were selected so that there were not many repeated values for any of the attributes (in principle, one would not expect repeated values for a continuous attribute, but there were repetitions in the data set).

Data set 2. $d = 13$ variables were drawn from the Commercial Building Energy Consumption Survey carried out by the Energy Information Administration of the U.S. Department of Energy. There are two categorical variables and 11 continuous variables. The number of records taken was 1080 (like for data set 1). Variables in this data set were more skewed than in data set 1.

A first round of experiments involved using rank swapping to mask the original data sets. Parameter values from $p = 1$ to $p = 20$ were considered. For each parameter choice, $Score'$ was computed using Equation (4). For each data set, Table 2 gives the parameter choices yielding the best $Score'$, the best IL', the best DLD' and the best ID'. For all measures, "best" means "lowest". Table 2 also gives the best values reached for each measure.

A second round of experiments used multivariate microaggregation taking three variables at a time (Mic3mul$k$) and four variables at a time (Mic4mul$k$), which were the best forms of microaggregation according to the score defined in [5]. Values for the parameter k between 3 and 18 were tried (k is the minimal size of microaggregates). For each data set, Table 3 gives the parameter choices

---

[1] http://www.census.gov/DES.

Table 2. Rank-swapping. Best parameter choices to minimize $Score'$, IL', DLD', and ID'

Data set			$Score'$	IL'	DLD'	ID'
1	Best parameter p		14	1	18	20
	Best measure value		25.663	1.95	12.355	29.541
2	Best parameter p		12	1	20	20
	Best measure value		23.687	2.037	11.839	29.022

yielding the best $Score'$, the best IL', the best DLD' and the best ID'. The best values reached for each measure are given as well.

Table 3. Multivariate microaggregation. Best parameter choices to minimize $Score'$, IL', DLD', and ID'

Data set	Method		$Score'$	IL'	DLD'	ID'
1	Mic3mul$k$	Best parameter k	14	3	18	17
		Best measure value	30.595	6.359	23.704	59.599
1	Mic4mul$k$	Best parameter k	14	3	17	18
		Best measure value	30.252	1.1	16.786	52.543
2	Mic3mul$k$	Best parameter k	14	3	18	17
		Best measure value	30.576	6.365	23.823	59.465
2	Mic4mul$k$	Best parameter k	17	3	16	18
		Best measure value	30.590	1.1	16.812	52.471

A third experimental round involved hybrid data computed using Algorithm 1 and an additive model (Equation (1)). The parameter here was α, which ranged between $\alpha = 0$ (pure LHS synthetic data) and $\alpha = 1$ (pure resampled original data). The size of the hybrid masked data set took values $n' = 500, 2000, 4000, 8000$. For each data set, Table 4 gives the parameter choices yielding the best $Score'$, the best IL', the best DLD' and the best ID'. The best values reached for each measure are given as well.

Note 1. An alternative to using simple random resampling with replacement in the first step of Algorithm 1 is to generate $n' - n$ LHS synthetic data records and replace each synthetic record with the nearest original record; the result is also a n'-record resampled original data set. Interesting as it may seem, this modification of the resampling procedure in Algorithm 1 does not lead to significant variation of the results shown in Table 4.

Note 2. For the data sets discussed in this section, the additive model of Equation (1) yields better results than the multiplicative model of Equation (2). However, this may change for other data sets.

Table 4. Additive LHS hybrid data. Best parameter choices to minimize $Score'$, IL', DLD', and ID'

Data set			$Score'$	IL'	DLD'	ID'
1	$n' = 500$	Best parameter α	0	1	0	0
		Best measure value	33.466	11.27	16.6	48.037
1	$n' = 2000$	Best parameter α	0	1	0	0
		Best measure value	32.069	5.363	19.336	50.775
1	$n' = 4000$	Best parameter α	0.1	1	0	0
		Best measure value	34.546	2.357	21.368	52.911
1	$n' = 8000$	Best parameter α	0	1	0	0
		Best measure value	38.854	1.982	23.964	55.139
2	$n' = 500$	Best parameter α	0.3	1	0	0
		Best measure value	34.225	7.504	18	46
2	$n' = 2000$	Best parameter α	0.1	1	0	0
		Best measure value	32.245	3.33	20	50.712
2	$n' = 4000$	Best parameter α	0.1	1	0	0
		Best measure value	31.940	6.929	21.614	52.827
2	$n' = 8000$	Best parameter α	0.4	1	0	0
		Best measure value	38.3	1.8	24.29	54.9

6   Conclusions

Looking at the performance in terms of $Score'$, it can be seen that rank swapping appears as the best performer, followed by multivariate microaggregation and then by LHS hybrid data (the difference between the last two approaches is small). The best scores of all methods are similar on both data sets, but the best parameter choice for hybrid data shows some dependency on the particular data set:

- For the first data set, LHS hybrid data score best when the data are purely synthetic ($\alpha = 0$) or with a weak resampled original component ($\alpha = 0.1$ for $n' = 2000$).
- For the second (more skewed) data set, a stronger resampled original component (α up to 0.4) may be needed to attain the best scores.

Regarding the global information loss IL', there is no clear winner. However, note that hybrid data get the lowest information loss when $\alpha = 1$, *i.e.* when the masked data are just resampled original data.

Note 3. If the IL'_1 component is suppressed from IL', then hybrid data tend to be the best performer (regardless of the parameter α): LHS nearly preserves averages, covariances and correlations, which brings IL_2, IL_3, IL_4 and IL_5 close to 0. In situations where IL'_1 is of critical importance, the LHS procedure should be performed at the subpopulation levels.

From the disclosure risk standpoint (DLD' and ID' measures), rank swapping is best, while hybrid data and multivariate microaggregation perform similarly.

Two additional lessons that can be learned from the tables in Section 5 are that:

- Mic3mul$k$ behaves similarly to Mic4mul$k$.
- Although there is no big influence of n' on the performance of additive hybrid data, experiments on both data sets show that taking $n' \approx 2n$ seems to be a wise option.

To summarize, best parameter choices for LHS synthetic microdata and multivariate microaggregation yield similar results, a bit behind those obtained with rank swapping.

References

1. R.A. Dandekar, "Performance improvement of restricted pairing algorithm for Latin hypercube sampling", in *ASA Summer Conference* (unpublished).
2. R.A. Dandekar, M. Cohen and N. Kirkendall, "Applicability of Latin hypercube sampling technique to create multivariate synthetic microdata", in *Proc. of ETK-NTTS 2001*. Luxembourg: Eurostat, pp. 839-847, 2001.
3. D. Defays and P. Nanopoulos, "Panels of enterprises and confidentiality: The small aggregates method", in *Proc. of 92 Symposium on Design and Analysis of Longitudinal Surveys*. Ottawa: Statistics Canada, 195-204, 1993.
4. J. Domingo-Ferrer, J.M. Mateo-Sanz, and V. Torra, "Comparing SDC methods for microdata on the basis of information loss and disclosure risk", *Proc. of ETK-NTTS 2001*. Luxemburg: Eurostat, pp. 807-825, 2001.
5. J. Domingo-Ferrer and V. Torra, "A quantitative comparison of disclosure control methods for microdata", in *Confidentiality, Disclosure and Data Access*, eds. P. Doyle, J. Lane, J. Theeuwes and L. Zayatz. Amsterdam: North-Holland, pp. 111-133, 2001.
6. J. Domingo-Ferrer and J.M. Mateo-Sanz, "Practical data-oriented microaggregation for statistical disclosure control", *IEEE Transactions on Knowledge and Data Engineering,* vol. 14, pp. 189-201, 2002.
7. R.L. Iman and W.J. Conover, "A distribution-free approach to inducing rank correlation among input variables", *Communications in Statistics*, vol. B11, no. 3, pp. 311-334, 1982.
8. M.A. Jaro, "Advances in record-linkage methodology as applied to matching the 1985 Census of Tampa, Florida", *Journal of the American Statistical Association*, vol. 84, pp. 414-420, 1989.
9. A.B. Kennickell, "Multiple imputation and disclosure protection: the case of the 1995 survey of consumer finances", in *Statistical Data Protection*. Luxembourg: Office for Official Publications of the European Communities, pp. 381-400, 1999.
10. M.D. McKay, W.J. Conover, and R.J. Beckman, "A comparison of three methods for selecting values of input variables in the analysis of output from a computer code", *Technometrics*, vol. 21, no. 2, pp. 239-245, 1979.
11. R. Moore, "Controlled data swapping techniques for masking public use microdata sets", U. S. Bureau of the Census, 1996 (unpublished manuscript).
12. L. Willenborg and T. de Waal, *Elements of Statistical Disclosure Control*. New York: Springer-Verlag, 2001.

Post-Masking Optimization of the Tradeoff between Information Loss and Disclosure Risk in Masked Microdata Sets*

Francesc Sebé[1], Josep Domingo-Ferrer[1],
Josep Maria Mateo-Sanz[2], and Vicenç Torra[3]

[1] Universitat Rovira i Virgili, Dept. of Computer Science and Mathematics
Av. Països Catalans 26, E-43007 Tarragona, Catalonia, Spain
{fsebe,jdomingo}@etse.urv.es
[2] Universitat Rovira i Virgili, Statistics and OR Group
Av. Països Catalans 26, E-43007 Tarragona, Catalonia, Spain
jmateo@etseq.urv.es
[3] Institut d'Investigació en Intel·ligència Artificial
Campus de Bellaterra, E-08193 Bellaterra, Catalonia, Spain
vtorra@iiia.csic.es

Abstract. Previous work by these authors has been directed to measuring the performance of microdata masking methods in terms of information loss and disclosure risk. Based on the proposed metrics, we show here how to improve the performance of any particular masking method. In particular, post-masking optimization is discussed for preserving as much as possible the moments of first and second order (and thus multivariate statistics) without increasing the disclosure risk. The technique proposed can also be used for synthetic microdata generation and can be extended to preservation of all moments up to m-th order, for any m.
Keywords: Microdata masking, Information loss, Disclosure risk, Post-masking optimization, Synthetic microdata generation.

1 Introduction

Statistical offices must guarantee statistical confidentiality when releasing data for public use. Statistical Disclosure Control (SDC) methods are used to to that end[7]. If data being released consist of individual respondent records, called microdata in the official statistics jargon, confidentiality translates to avoiding disclosure of the identity of the individual respondent associated with a published record. At the same time, SDC should preserve the informational content to the maximum extent possible. SDC methods are an intermediate option between encryption of the original data set (no disclosure risk but no informational content released) and straightforward release of the original data set (no confidentiality

---

* Work partly supported by the European Commission under project IST-2000-25069 "CASC".

J. Domingo-Ferrer (Ed.): Inference Control in Statistical Databases, LNCS 2316, pp. 163–171, 2002.

but maximal informational content released). SDC methods for microdata are also known as masking methods.

In [2,3], a comparison of masking methods for microdata was conducted. The comparison was done by computing a score which weighed information loss against disclosure risk between the original and the masked microdata set. In [1], the score was modified to cope with the case where the number of records in the original and the masked data sets is not the same.

In this paper, we present a post-masking optimization procedure which seeks to modify the masked data set to preserve the first and second-order moments of the original data set as much as possible without increasing the disclosure risk. The better the first and second-order moments are preserved, the better will multivariate statistics on the resulting masked data set mimic those that would be obtained on the original data set. In order to avoid substantially increasing the disclosure risk, a constraint is imposed to guarantee that individual data resulting from post-masking optimization are not too similar to individual original data.

The optimization procedure presented can be combined with any masking method for numerical microdata and can lead to an improvement in terms of the score defined in [3]: the reason is that the information loss is reduced without significantly increasing the disclosure risk. The procedure can also be used in a stand-alone way to produce synthetic data sets with prescribed first and second-order moments. Furthermore, an extension aiming at preservation of all moments up to m-th order for any m is straightforward.

Section 2 sketches the score constructed in [3] with the modifications of [1]. Section 3 describes the post-masking optimization problem and a heuristic hill-climbing procedure to solve it. Computational results reflecting improvement of rankswapped data and microaggregated data are presented in Section 4. Section 5 presents some conclusions and sketches two extensions: 1) how to use the proposed optimization as a synthetic microdata generator; 2) how to preserve all moments up to m-th order for any m.

2   A Score for Method Comparison

Let n be the number of records in the original data set and n' the number of records in the masked data set. Let d be the number of variables (assumed to be the same in both data sets). Let X be an $n \times d$ matrix representing the original data set: columns correspond to variables and rows correspond to records. Similarly, let X' be an $n' \times d$ matrix representing the masked data set. Let V and V' be the covariance matrices of X and X', respectively; similarly, let R and R' be the correlation matrices. Let $\bar{X}$ and $\bar{X}'$ be the vectors of averages of variables in X and X'. Finally, let S and S' be the vectors of variances of variables in X and X'. Define the mean absolute error of a matrix A' vs another matrix A as the average of the absolute values of differences of corresponding components in both matrices (what "corresponding" means will be discussed below); define the mean variation as the average of absolute differences of corresponding components in both matrices with each difference divided by the absolute value of the component in A.

Disclosure risk can be measured using record linkage. Two record linkage methods were used in [3]:

Distance-based record linkage. Let the original and masked data sets consist both of d variables (it is assumed that both data sets contain the same variables). We define that a record in the masked data set corresponds to the nearest record in the original data set, where "nearest" means at shortest d-dimensional Euclidean distance. Assume further that the intruder can only access i key variables of the original data set and tries to link original and masked record based on these i variables. Linkage then proceeds by computing i-dimensional distances between records in the original and the masked data sets (distances are computed using only the i key variables) The variables used are standardized to avoid scaling problems. A record in the masked data set is labeled as "correctly linked" when the nearest record using i-dimensional distance is the corresponding one (*i.e.* the nearest record using d-dimensional distance).

Probabilistic record linkage. Defined in [5], uses a matching algorithm to pair records in the masked and original data sets. The matching algorithm is based on the linear sum assignment model. The definition of "correctly linked" records is the same as in distance-based record linkage. This method is attractive because it only requires the user to provide two probabilities as input: one is an upper bound of the probability of a false match and the other an upper bound of the probability of a false non-match. Unlike distance-based record linkage, probabilistic record linkage does not require rescaling variables nor makes any assumption on their relative weight (by default, distance-based record linkage assumes that all variables have the same weight). Furthermore, probabilistic record linkage can be used on both numerical and categorical data.

A score combining information loss measures with disclosure risk measures can be constructed as follows:

IL Information Loss: 100 times the average of the mean variation of X' vs X (called IL_1), the mean variation of $\bar{X}'$ vs $\bar{X}$ (called IL_2), the mean variation of V' vs V (called IL_3), the mean variation of S' vs S (called IL_4) and the mean absolute error of R' vs R (called IL_5).

DLD Average of DLD-1, $\cdots$, DLD-7. DLD-i is the percent of records correctly linked using distance-based record linkage with Euclidean distance when the intruder knows i key variables of the original file. See below for a discussion on what "correctly linked" means.

PLD Same as DLD, but for probabilistic record linkage.

ID Average percent of original values falling in the intervals around their corresponding masked values. The average is over interval widths from 1% to 10%.

Overall score

$$Overall\_score = 0.5 \cdot IL + 0.125 \cdot DLD + 0.125 \cdot PLD + 0.25 \cdot ID \quad (1)$$

A simplified version of the score can be used with only one record linkage method, namely DLD. In this case, the score is

$$Score = 0.5 \cdot IL + 0.25 \cdot DLD + 0.25 \cdot ID \quad (2)$$

The lower *Score*, the better is a method. In computation of IL_2, IL_3, IL_4 and IL_5 "corresponding" components between the matrices for the original and masked data sets means "referring to the same variables". To compute IL_1 and ID, we need to define a correspondence between records in X and X'. A natural way is to map each published masked record i to the nearest original record $c(i)$, using the d-dimensional Euclidean distance between records (where d is the number of variables in the data sets).

Also, computation of the disclosure risk measure DLD (and PLD) requires defining what correct linkage means. If the masked and the original data sets have d variables, we say that a masked record is correctly linked to an original record if they are at the shortest possible d-dimensional Euclidean distance.

3 Post-Masking Optimization

Once an original data set X has been masked as X', post-masking optimization aims at modifying X' into X'' so that the first and second-order moments of X are preserved as much as possible by X'' while keeping IL_1 around a prescribed value. Near preservation of first and second-order moments results in (constrained) minimization of IL_2, IL_3, IL_4 and IL_5. Regarding IL_1, we cannot pretend to minimize it, because disclosure risk would most likely suffer a dramatic increase: post-masking optimized data would look too much like the original data.

3.1 The Model

The first-order moments of X depend on the sums

$$\frac{\sum_{i=1}^{n} x_{ij}}{n} \text{ for } j = 1, \cdots, d \quad (3)$$

where x_{ij} is the value taken by the j-th variable for the i-th record. The second-order moments of X depend on the sums

$$\frac{\sum_{i=1}^{n} x_{ij}^2}{n} \text{ for } j = 1, \cdots, d \quad (4)$$

$$\frac{\sum_{i=1}^{n} x_{ij} x_{ik}}{n} \text{ for } j, k = 1, \cdots, d \text{ and } j < k \quad (5)$$

Therefore, our goal is to modify X' to obtain a X'' so that the above $2d + d(d - 1)/2$ sums are nearly preserved while keeping IL_1 and disclosure risk similar in X' and X''. First, let us compute IL_1 of X' vs X as

$$IL_1 := 100 \left(\frac{\sum_{i=1}^{n'} \sum_{j=1}^{d} \frac{|x'_{ij} - x_{c(i),j}|}{|x_{c(i),j}|}}{dn'} \right) \tag{6}$$

where $c(i)$ is the original record nearest to the i-th masked record of X' (d-dimensional Euclidean distance is used). Now let $0 < q \leq 1$ be a parameter and let M be the set formed by the $100q\%$ records of X' contributing most to IL_1 above. Then let us compute the values x''_{ij} of X'' as follows. For $x'_{ij} \notin M$ then $x''_{ij} := x'_{ij}$. For $x'_{ij} \in M$ the corresponding x''_{ij} are solutions to the following minimization problem:

$$\min_{\{x''_{ij} | x'_{ij} \in M\}} \sum_{j=1}^{d} \left(\frac{\sum_{i=1}^{n'} x''_{ij}}{n'} - \frac{\sum_{i=1}^{n} x_{ij}}{n} \right)^2 + \sum_{j=1}^{d} \left(\frac{\sum_{i=1}^{n'} x''^2_{ij}}{n'} - \frac{\sum_{i=1}^{n} x^2_{ij}}{n} \right)^2$$

$$+ \sum_{1 \leq j < k \leq d} \left(\frac{\sum_{i=1}^{n'} x''_{ij} x''_{ik}}{n'} - \frac{\sum_{i=1}^{n} x_{ij} x_{ik}}{n} \right)^2 \tag{7}$$

subject to

$$0.99 \cdot p \cdot IL_1 \leq \frac{\sum_{j=1}^{d} \sum_{i=1}^{n'} \frac{|x''_{ij} - x_{C(i),j}|}{|x_{C(i),j}|}}{dn'} \leq 1.01 \cdot p \cdot IL_1 \tag{8}$$

where $p > 0$ is a parameter and $C(i)$ is the original record nearest to the i-th masked record of X'' after optimization. Note that, in general, $C(i) \neq c(i)$, because in general $X'' \neq X'$.

3.2   A Heuristic Optimization Procedure

To solve the minimization problem (7) subject to constraint (8), the following hill-climbing heuristic procedure has been devised:

Algorithm 1 (PostMaskOptim(X,X',p,q,TargetE)).

1. *Standardize all variables in X and X' by using for both data sets the averages and standard deviations of variables in X.*
2. *Compute IL_1 between X and X' according to expression (6).*
3. *Let $TargetIL_1 := p \cdot IL_1$.*
4. *Let $X'' := X'$.*
5. *Rank records in X'' according to their contribution to IL_1. Let M be the subset of the $100q\%$ records in X'' contributing most to IL_1.*
6. *For each record i in X'', determine its nearest record $C(i)$ in X (use d-dimensional Euclidean distance).*

7. *Compute E, where E denotes the objective function in Expression (7).*
8. *While $E \geq TargetE$*
   (a) *Randomly select one value v of a record i_v in $M \subset X''$ and randomly perturb it to get v'. Replace v with v' in record i_v.*
   (b) *Recompute the nearest record $C(i_v)$ in X nearest to the updated i_v.*
   (c) *Let $PreviousIL_1 := IL_1$.*
   (d) *Compute IL_1 between X and X''. To do this, use expression (6) while replacing x'_{ij} by x''_{ij} and $c(i)$ by $C(i)$.*
   (e) *Let $PreviousE := E$.*
   (f) *Recompute E $(X''$ has been modified).*
   (g) *If $E \geq previousE$ then undo := true.*
   (h) *If $IL_1 \notin [0.99 \cdot TargetIL_1, 1.01 \cdot TargetIL_1]$ and $|IL_1 - TargetIL_1| \geq |PreviousIL_1 - TargetIL_1|$ then undo := true.*
   (i) *If undo = true then restore the original value v of record i_v and recompute the nearest record $C(i_v)$ in X nearest to i_v.*
9. *Destandardize all variables in X and X'' by using the same averages and standard deviations used in Step 1.*

Note that, by minimizing E, the algorithm above attempts to minimize the information loss IL. No direct action is taken to reduce or control disclosure risk measures DLD and ID, beyond forcing that IL_1 should be in a pre-specified interval to prevent the optimized data set from being dangerously close to the original one. The performance of Algorithm 1 is evaluated *a posteriori*: once E reaches $TargetE$, the algorithm stops and yields an optimized data set for which IL, DLD and ID must be measured.

4   Computational Results

The test microdata set no. 1 of [1] was used. This microdata set was constructed using the Data Extraction System (DES) of the U.S. Census Bureau (http://www.census.gov/DES). $d = 13$ continuous variables were chosen and 1080 records were selected so that there were not many repeated values for any of the attributes (in principle, one would not expect repeated values for a continuous attribute, but there were repetitions in the data set).

In the comparison of [2,3], two masking methods were singled out as particularly well-performing to protect numerical microdata: rank swapping [6] and multivariate microaggregation [4]. For both methods, the number of masked records is the same as the number of original records ($n = n' = 1080$). Several experiments have been conducted to demonstrate the usefulness of post-masking optimization to improve on the best (lowest) scores reached by rank swapping and multivariate microaggregation.

The first row of Table 1 shows the lowest score reached by rank swapping for the test microdata set: the score is 25.66 and is reached for parameter value 14 (see [1]). The next rows of the table show scores reached when Algorithm 1 is used with several different values of parameters p (proportion between target IL_1 and initial IL_1) and q (proportion of records in M). The last column shows the value of the objective function E reached (for all rows but the first one, this is the $TargetE$ parameter of Algorithm 1). The score is computed using Expression (2) and the values of IL, DLD and ID reached are also given in Table 1.

Table 1. Rank-swapping with parameter 14. First row, best score without optimization; next rows, scores after optimization

p	q	$Score$	IL	DLD	ID	E
None	None	25.66	23.83	14.74	40.23	0.419
0.5	0.5	24.45	14.73	20.30	48.03	0.04
0.5	0.3	22.15	13.65	16.30	44.98	0.04
0.5	0.1	21.71	15.26	14.81	41.51	0.09

The first row of Table 2 shows the lowest score reached by multivariate microaggregation for the test data set: the score is 31.86 and is reached for parameter values 4 and 10, that is, when four variables are microaggregated at a time and a minimal group size of 10 is considered (see [1]). The next rows of the table show scores reached when Algorithm 1 is used with several different values of parameters p and q.

Table 2. Multivariate microaggregation with parameters 4 and 10. First row, best score without optimization; next rows, scores after optimization

p	q	$Score$	IL	DLD	ID	E
None	None	31.86	22.48	22.14	60.34	0.122
0.5	0.5	26.96	14.16	21.06	58.54	0.008
0.5	0.3	27.39	14.74	21.29	58.80	0.008
0.5	0.1	28.03	14.94	21.83	60.38	0.008

When looking at the results on rankswapped data (Table 1), we can observe the following:

- There is substantial improvement of the score: 21.71 for post-masking optimization with $p = 0.5$ and $q = 0.1$ in front of 25.66 for the initial rankswapped data set.
- The lower q (*i.e.* the smaller the number of records altered by post-masking optimization), the better is the score. In fact, the score for $q = 0.1$ is lower than for $q = 0.3, 0.5$ even if the target E for $q = 0.1$ is less stringent (higher) than for the other values of q.
- Post-masking optimization improves the score by reducing information loss IL and hoping that disclosure risks DLD and ID will not grow. In fact,

Table 1 shows that DLD and ID increase in the optimized data set with respect to the rankswapped initial data set. The lower q, the lower is the impact on the rankswapped initial data set, which results in a smaller increase in the disclosure risk. This small increase in disclosure risk is dominated by the decrease in information loss, hence the improved score.

The results on microaggregated data (Table 2) are somewhat different. The following comments are in order:

- Like for rankswapping, there is substantial improvement of the score: 26.96 for post-masking optimization with $p = 0.5$ and $q = 0.5$ in front of 31.86 for the initial microaggregated data set.
- The higher q, the better is the score. This can be explained by looking at the variation of IL, DLD and ID. Microaggregated data are such that there is room for decreasing IL while keeping DLD and ID at the same level they had in the initial microaggregated data set. In this respect, we could interpret that, multivariate microaggregation being "less optimal" than rank swapping, we should not be afraid of changing a substantial number of values because this can still lead to improvement.

5   Conclusions and Extensions

The procedure presented here is designed to minimize information loss between a masked data set and its original version. Although disclosure risk is not explicitly considered in the minimization model described in Subsection 3.1, there is a constraint on IL_1 whose purpose is to prevent the optimized masked data set from being too close to the original one.

The described post-masking optimization can be applied to improve any masking method. We have demonstrated improvement in the case of two microdata masking methods which already had been identified as the best performers for numerical microdata protection. This is a substantial step forward in optimizing the tradeoff between information loss and disclosure risk in microdata protection.

The application of the proposed technique can be extended in at least two directions:

- *Synthetic microdata generation.* Algorithm 1 can be used on a stand-alone basis to generate synthetic microdata. To do this, let input parameter p be small, let $q := 1$ and let input parameter X' be a random data set with the same number of variables as the original data set X. The resulting synthetic data set will be such that the objective function E reaches the pre-specified value $TargetE$ and IL_1 is $p\%$ of the initial (big) IL_1.
- *Preservation of all moments up to m-th order.* Given a positive integer m, all what is needed to preserve all moments up to m-th order is to add to the objective function (7) terms corresponding to the squared differences between the i-th order sums of X and X'', for $i = 1$ to m.

References

1. R.A. Dandekar, J. Domingo-Ferrer, and F. Sebé, "LHS-based hybrid microdata vs rank swapping and microaggregation for numeric microdata protection", in *Inference Control in Statistical Databases*, LNCS 2316, Springer 2002, pp. 153–162.

2. J. Domingo-Ferrer, J.M. Mateo-Sanz, and V. Torra, "Comparing SDC methods for microdata on the basis of information loss and disclosure risk", *Proc. of ETK-NTTS 2001*. Luxemburg: Eurostat, pp. 807–825, 2001.

3. J. Domingo-Ferrer and V. Torra, "A quantitative comparison of disclosure control methods for microdata", in *Confidentiality, Disclosure and Data Access*, eds. P. Doyle, J. Lane, J. Theeuwes and L. Zayatz. Amsterdam: North-Holland, pp. 111-133, 2001.

4. J. Domingo-Ferrer and J.M. Mateo-Sanz, "Practical data-oriented microaggregation for statistical disclosure control", *IEEE Transactions on Knowledge and Data Engineering,* vol. 14, pp. 189–201, 2002.

5. M.A. Jaro, "Advances in record-linkage methodology as applied to matching the 1985 Census of Tampa, Florida", *Journal of the American Statistical Association,* vol. 84, pp. 414–420, 1989.

6. R. Moore, "Controlled data swapping techniques for masking public use microdata sets", U. S. Bureau of the Census, 1996 (unpublished manuscript).

7. L. Willenborg and T. de Waal, *Elements of Statistical Disclosure Control.* New York: Springer-Verlag, 2001.

The CASC Project

Anco Hundepool

Statistics Netherlands
Methods and Informatics Department, P.O. Box 4000
2270 JM Voorburg, The Netherlands
ahnl@krypton.vb.cbs.nl

Abstract. In this paper we will give an overview of the 5th framework CASC (Computational Aspects of Statistical Confidentiality) project. This project can be seen as a follow up of the 4th Framework SDC-project. However, the main emphasis is more on building practical tools. The further development of the ARGUS-software will play a central role in this project. Besides this software development, several research topics have been included in the CASC-project. These research topics, both for the disclosure control of microdata as well as tabular data, aim at obtaining practical results that might be implemented in future version of ARGUS and find its way to the end-users.
Keywords: Statistical Disclosure Control, μ-ARGUS, τ-ARGUS, microdata, tabular data.

1   Introduction

Statistical Disclosure Control is a field in statistics that has attracted much attention in recent years. Decision-makers demand more and more detailed statistical information. Researchers at universities and similar institutes have the capacity to perform complex statistical analysis on their powerful PCs and they desire detailed microdata. Therefore there is a growing pressure on the statistical offices to publish more and more detailed information. The other side however is that statistical offices have a legal or moral obligation to protect the confidentiality of information provided to them by respondents. This confidentiality is vital also to guarantee the future co-operation of respondents.

This puts a large responsibility on the shoulders of statistical offices to minimise the risks of disclosure from the information that they make available from their censuses and surveys. The question then arises how the information available can be modified in such a way that the data released can be considered statistically useful and do not jeopardise the privacy of the entities concerned. The aim of Statistical Disclosure Control (SDC) is to diminish the risk that sensitive information about or from individual respondents can be disclosed from a data set. The data set can be either a microdata set or a table. A microdata set consists of a set of records containing information on individual respondents or economic entities. A table contains aggregate information of individual entities.

The CASC project on the one hand can be seen as a follow up of the SDC-project of the 4th Framework. It will build further on the achievements of that successful

J. Domingo-Ferrer (Ed.): Inference Control in Statistical Databases, LNCS 2316, pp. 172 - 180, 2002.

project. On the other hand it will have new objectives. It will concentrate more on practical tools and the research needed to develop them. For this purpose a new consortium has been brought together. It will take over the results and products emerging from the SDC-project. One of the main tasks of this new consortium will be to further develop the ARGUS-software, which has been put in the public domain by the SDC-project consortium and is therefore available for this consortium. The main software developments in CASC are μ-ARGUS, the software package for the disclosure control of microdata while τ-ARGUS handles tabular data.

The CASC-project will involve both research and software development. As far as research is concerned the project will concentrate on those areas that can be expected to result in practical solutions, which can then be built into (future version of) the software. Therefore the CASC-project has been designed round this software twin ARGUS. This will make the outcome of the research readily available for application in the daily practice of the statistical institutes.

At this workshop will see several presentations of parts of the CASC-project we will give only a general overview of the project leaving the details to these other CASC-presentations.

2   CASC-Partners

At first sight the CASC-project team had become rather large. However there is a clear structure in the project, defining which partners are working together for which tasks. Sometimes groups working closely together have been split into independent partners only for administrative reasons.

Institute	Short	Country
- Statistics Netherlands	CBS	NL
- Istituto Nazionale di Statistica	ISTAT	I
- University of Plymouth	UoP	UK
- Office for National Statistics	ONS	UK
- University of Southampton	SOTON	UK
- The Victoria University of Manchester	UNIMAN	UK
- Statistisches Bundesamt	StBA	D
- University La Laguna	ULL	ES
- Institut d'Estadistica de Catalunya	IDESCAT	ES
- Institut National de Estadística	INE	ES
- TU Ilmenau	TUIlm	D

	Institut d'Investigació en Intelligència Artificial-CSIC	CIS	ES
-	Universitat Rovira i Virgili	URV	ES
-	Universitat Politècnica de Catalunya	UPC	ES

Although Statistics Netherlands is the main contractor, the management of this project is a joint responsibility of the steering committee. This steering committee constitutes of 5 partners, representing the 5 countries involved and also bearing a responsibility for a specific part of the CASC-project:

CASC Steering Committee

Institute	Country	Responsibility
Statistics Netherlands	Netherlands	Overall manager Software development
Istituto Nationale di Statistica	Italy	Testing
Office for National Statistics	UK	
Statistisches Bundesamt	Germany	Tabular data
Universitat Rovira i Virgili	Spain	Microdata

3   ARGUS Software Development

3.1 Software Concepts

As the CASC-project aims at practical solutions for disclosure control, we have given the development of the ARGUS software a central role in the project. The ARGUS software will play the binding factor between the different parts of the project. Research topics have only been included if they aim at results that either can be implemented in (future) version of ARGUS or aim at testing the methodology used in the CASC-project.

The starting point for the CASC-project were the ARGUS-twins resulting from the SDC-project. However as these twins had been developed with Borland C++, we have decided to convert the software to a more modern, up-to-date version of C++, i.e. Visual C++. However for the user-interfaces we use Visual Basic as a programming tool. This is an easier platform for the development of user-interfaces, still meeting the needs of ARGUS. For the more crucial routines taking care of the heavy calculations we use Visual C++, which will lead to more efficient code. Some methods in ARGUS lead to complex computation problems, which justifies the choice for C++.

The routines build in Visual C++ will be compiled into an OCX-component, which can easily be used in the Visual Basic user-interface program. This guarantees a more flexible software concept and gives better options for the inclusion of additional routines for disclosure control even third party solutions. A first example is the link be-tween ARGUS and the German GHQUAR/GHMITER software. However the aims with ARGUS in the CASC project are that ARGUS should be expanded into a control centre that will offer the user to choice of SDC-solutions. This also makes the com-parison between the different solutions within one framework much easier.

3.2 μ-ARGUS

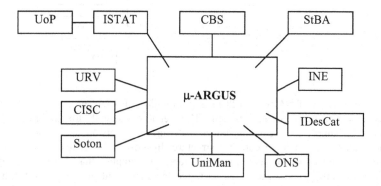

Although the current version of ARGUS (3.0) has adequate routines for the disclosure protection of social micro data, we lack solutions for economic micro data. This is due to different characteristics of these data. The distribution of the variables is much more skewed, making it much easier to re-identify the individual records. Also the legal restrictions in many countries as much more restrictive for the access of economic micro data. The following enhancements to ARGUS software will provide techniques and methods not yet available to SDC practitioners, and will be a major step forward in its field.

- Micro-aggregation has been proved to be a promising method. Studies by Josep Domingo support this and the inclusion of micro-aggregation routines is planned.
- PRAM (Post RAndoMisation) is a technique under investigation a.o. at Statistics Netherlands. The basic idea is to add noise to the data. The distribution of the noise will also be made available to the potential data users. The combination of the distorted data and the known noise allows the user to make good estimates of population table etc. This might be a drawback for the inexperienced users, but he advantage is that much more detailed information can be released. Good additional (tabulation) software could overcome this.
- Matrix Masking. Masking techniques have been studied already for some time, but we think that the timeis ready for the inclusion in μ-ARGUS.
- Disclosure risk models will be implemented. Based on the work of Skinner, The Italian partner, Luisa Franconi, has prepared a practical implementation for ARGUS. This gives a more sophisticated approach for the estimation of the disclosure risk of the individual records.

3.3 τ-ARGUS

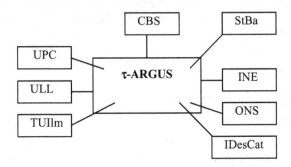

The current version of τ-ARGUS (2.0) has been build to implement the optimisation approach by JJ Salazar and M. Fischetti. This version is suitable for the disclosure protection of simple 3-dimensional tables. Although this version works nice, most practical tables have a more complex structure. In most tables at least one of the spanning variables have a hierarchical structure. This implies that there are many more sub-totals that makes it much easier to recalculate individual cells. Therefor the mathematical optimisation problems are much more complex.

We will include novel solutions to the particularly large and complex optimisation problems associated with SDC, which will allow minimal loss of information from treatment of multi-dimensional tables, and will use algorithms based on state of the art programming tools. Solving these problems is a challenging task, for which two groups of mathematical programming experts have joined the CASC team.

The main approach will be by Fischetti/Salazar by extending the current library with facilities allowing for more complex data structures. Jordi Castro will investigate the use of networks. Currently only solutions based on networks are known for two-dimensional tables, but research is carried out to extend this.

However it is to expected that these complex mathematical solutions will work fine for small and medium sized problems, but we also need solutions for very large tables.

For that extend we have made a method, HITAS, which uses the current available solutions for unstructured tables. The large structured table is then broken down into many small unstructured tables and protected. By applying this method top-down we end up with a protection of the whole table. Sometimes some backtracking is necessary, but that problem is taken care of. In this way rather large tables can be protected.

A second solution for very big and complex tables is provided by the German GHQUAR/GHMITER program. This method based on hypercubes can solve tables with hierarchies up to 7 dimensions and also allows for linked tables. The drawback can be that the solution provided will be over-protective, because the problem is not solved to optimality. However this will at least open a solution for the very big tables, which – for the time being- cannot be solved optimally.

Besides this we aim at extending τ-ARGUS into a control-centre for tabular disclosure control, making other techniques (new and existing) more easily available through τ-ARGUS. This will offer the users a range of solutions with all its own qualities and characteristics. The user can much more easy compare the different solutions, had the choice for a quick but not optimal solutions or more slow, but optimal solutions. Also we can now also investigate the drawbacks of the non-optimal solutions.

4   Methodology Research for Microdata

4.1   Introduction

It is foreseen in the CASC-project that several new techniques for disclosure protection will be implemented. The need for these new techniques lies in the fact that the currently used methods like global recoding and local suppression serve very well the needs for social survey data but are inadequate for the disclosure protection of business microdata. New techniques investigated are micro-aggregation, noise addition, PRAM (Post-randomisation) and masking techniques. The research on PRAM is formally not part of the CASC project as this research is being carried out already as a PhD research at Statistics Netherlands. However, the results will be implemented in μ-ARGUS. Noise addition and masking techniques are studied and a special study into an alternative method for business data preserving the individual profile for each unit will be undertaken. Micro-aggregation will be studied as an alternative.

In addition to these new techniques for disclosure protection risk models will be investigated. These disclosure risk models help to assess the safety of a protected microdata file. A study on record level measures will result in a research report on noise addition. These latter will result in research that might be implemented in ARGUS during the CASC-project, but will be implemented only after the foreseen scope of this project.

A simulation of the intruder will be investigated, when attempts will be made to undo the disclosure protection. An other important study is into the effects on the analytical power of the protected microdata file, i.e. how well are these protected microdata files suited for statistic analysis projects.

The different approaches for this topic are justified by the need for safe business microdata files, for which few solutions are available. The implementation of these methods in ARGUS will allow for an easy application of these methods, which will result in growing insight in the quality and the applicability of these methods. In the long run we might reach a common opinion on recommendations for the generation of safe business microdata files. Eventually this offers the possibility of European harmonisation.

4.2 Methodology for Business Microdata

Research in this area is at an early stage, with however some applications successfully attempted. The project work will be focused on the practical need for users to have a secure methodological framework within which they can select suitable techniques to effectively treat small to medium size business microdata. Research topics include:

- Building of a new framework for business microdata that will maintain an individual profile for each unit.
- Development of matrix masking methods to allow their application to the complex data structures found in practice
- Further refinement of microaggregation techniques

4.3 Measurement of Risk and Information Loss

A project aim is to incorporate realistic measures of risk, and when they become available measures of information loss, into the ARGUS software to make available to users. Users will then for the first time have the tools to make a properly informed choice between different methods of SDC treatment, which will balance risk of disclosure against cost in terms of information loss.

- Extension of record level measures to take account of the possible misclassification of key variables and of emerging ideas on record-linkage
- Setting a framework to work towards measures of information loss, with a first attempt to quantify the loss.
- A feature of elements of this proposal will be the incorporation of the measurement of risk and of loss of analytical validity into research on data perturbation techniques.

5 Methodology for Tabular Data

5.1 Introduction

τ-ARGUS for tabular data resulting from the SDC-project covers the disclosure protection of simple unstructured tables up to dimension 3. A central role in the disclosure protection of tables is played by the dominance rule. This rule states which cells in a table are unsafe and therefore cannot be published. Alternatives for the dominance rule (the pq-rule) will be made available as well.

Due to the presence of marginals in a table it is often easy to recalculate these suppressed cells. So additional cells must be suppressed to prevent this recalculation of the primary unsafe cells. It is not only enough to prevent exact recalculation but also to guarantee a safety range to protect the primary unsafe cells. The optimal selection of these secondary cells, as to avoid unnecessary high losses in the information content of the protected tables, is a very complex numerical optimisation problem.

Although in the τ-ARGUS-version resulting from the SDC-project a solution is available for unstructured tables, it cannot be applied to many tables in the daily life of a statistical office, because they have a hierarchical structure. These hierarchical structures imply many more (sub-)marginals, which can be used to recalculate these primary suppressed cells. Also the linked tables, having some marginals in common, must be treated simultaneously.

This makes the optimisation problem to find the optimum suppression pattern still much harder. Even for renowned researchers in the field of numerical optimisation this is a very hard problem. Nevertheless we aim at a solution for this hard problem. The main approach is undertaken by JJ Salazar, dealing with the research required to specify the new models before implementation and testing. A second supporting approach is based on network flow algorithms.

Besides these complex optimisation approaches we will develop and implement heuristic methods, which aim at a much quicker solution. It is also to be expected that these methods will be able to solve much larger instances. The price for this will however be a non-optimal solution. It is known from previous investigations that τ-ARGUS is able to reduce the information loss for about 30 to 50%. For several tables this advantage of speed might prove to be adequate. Some of these methods are already available in a basic form (e.g. GHQUAR) but we will extend τ-ARGUS to facilitate the access to these heuristic methods. Another approach is based on the non-hierarchical solutions already available, by breaking down the big hierarchical table into several sub-problems.

One of the outcomes of this project is the composition of a set of test-tables. These tables will play the role of test-bench for the optimisation procedures and are of vital value for the researchers in numerical optimisation techniques to find the best solutions.

5.2   Main Objectives in Tabular Data Research

The three main goals of innovation in the proposed project regarding tools for tabular data protection will be:

- Firstly to develop data-structures for τ-ARGUS that are able to represent the cell suppression problem for hierarchical structured and linked tables.
- Secondly, GHQUAR will be integrated into the restructured version of τ-ARGUS.
- The third main task will be to speed up the linear programming methodology in ARGUS as emerged from the 4th Framework project. This will make τ-ARGUS capable of solving the larger problems that result from the representation of real life tables with many sub-marginals in reasonable (computing) time.

It should be noted, assuming that the computational burden was not an issue at all, a straightforward change of the data-structure would do, to make the current linear programming approach applicable to hierarchical structured and linked tables too. However, in real life the computational burden is an issue indeed, making it quite a challenge to preserve the excellent performance of the linear programming approach with respect to information loss, while speeding it up sufficiently for moderate to larger sized applications. (It won't certainly be possible to bring it to the extremes, e.g. make it applicable to those X-large applications, that can still be handled efficiently by GHQUAR).

6   Testing

Under this chapter we make a distinction between the actual testing of the software and the testing of the methodology. It is very important to see how much protection is gained from the methods implemented. So one of our partners, Mark Eliot, is playing the role of simulating the intruder. This work can be seen as a test-case of the (newly) developed methodology.

The actual testing of the software gets serious attention too. Lessons drawn from its predecessor, the SDC project, have learnt that you cannot only rely on voluntary testers. Therefore testing has been incorporated as a separate task in the project. Both the building of test-sets as well as the actual testing of the software tools is an essential part of the CASC-project.

7   Conclusion

The major objective of this project is that the results will be used in real life situations in official statistics. The composition of the project team has been designed in such a way that the primary users, i.e. the NSI's, are active members. Seven statistical offices (5 national and two regional) participate in the project, either actively in the various stages of the development or as testers of the results. This reflects the needs and the interest of the NSI's for these kinds of tools.

Side effects of this project will be that the research community on Statistical Disclosure Control in Europe will work together. This joint effort will bring the state of the art to a higher level.

In order to disseminate the results of the CASC-project the project team will maintain a WEB-site (http://neon.vb.cbs.nl/casc). Research papers resulting from this project as well as other material of interest for this field will find a place there.

Tools and Strategies to Protect Multiple Tables with the GHQUAR Cell Suppression Engine

Sarah Giessing[1] and Dietz Repsilber[2]

[1]Federal Statistical Office of Germany
65180 Wiesbaden
Sarah.giessing@statistik-bund.de
[2]Landesamt für Datenverarbeitung und Statistik Nordrhein-Westfalen
Mauerstrasse 51, 40476 Düsseldorf
Dietz.Repsilber@LDS.NRW.DE

Abstract. The paper will introduce into the method of the GHQUAR algorithm and will provide information on how to use the software. Recently, a procedure to protect sets of multiple linked tables has become available for GHQUAR. This procedure, the software GHMITER, will also be illustrated in this paper. Another new tool to support GHQUAR which is needed particularly when the program is used to protect tables in a user-demand driven tabulation process will be presented as well. Finally, the paper gives some information on plans for the development of graphical user interfaces for GHQUAR and the supporting routines, relating particularly to the link between τ–ARGUS and GHQUAR to be created as one task of the CASC project.

1 Introduction

Data collected within government statistical systems usually must be provided as to fulfill requirements of many users differing widely in the particular interest they take in the data. Statisticians try to cope with this range of interest in their data, providing the data at several levels of detail in large tables, based on elaborate hierarchical classification schemes. Usually, some of the cells of such a statistical table must be kept confidential to avoid disclosure of individual respondent data. All the linear relationship between the cells of such a table must be considered when selecting a suitable pattern of suppressed cells. The problem of finding an optimum set of suppressions is known as the 'secondary cell suppression problem'. It is computationally extremely hard to find exact, or close-to-optimum solutions for the secondary cell suppression problem for large hierarchical tables.

The GHQUAR algorithm developed at the Statistical office of Northrhine-Westphalia/Germany offers a quick heuristic solution (c.f. [4]). Together with supporting routines that have become available recently, it is possible to protect tables of any size and complexity using the GHQUAR method.

J. Domingo-Ferrer (Ed.): Inference Control in Statistical Databases, LNCS 2316, pp. 181 - 192, 2002.

It is one of the tasks of the EPROS[1] project CASC (Computational Aspects of Statistical Confidentiality) to extend the software τ-ARGUS for tabular data protection into a control-center for tabular-data protection, particularly making GHQUAR and the above mentioned interface procedures more easily available through τ-ARGUS [8].

This paper will introduce into the method of GHQUAR (Section 2), focussing particularly on the underlying concept of interval protection and information loss, and give background information relating to certain control options of the system. We will further illustrate the method for table-to-table protection as implemented in the system GHMITER (Section 3) and describe facilities of the tool POOLAC supposed to be used to co-ordinate suppression patterns in multiple linked tables (Section 4).

Section 5 will focus on the plans for graphical user interfaces for the GHQUAR algorithm, in particular on plans for the CASC project, but will also comment on an existing user interface developed on behalf of Eurostat.

We assume that the reader is acquainted with the concepts of secondary cell suppression, otherwise we refer to, for instance, [7] for a general introduction.

2   The GHQUAR Algorithm for Secondary Cell Suppression

At the Landesamt für Datenverarbeitung und Statistik in Nordrhein-West-falen/Germany a hypercube method for secondary cell suppression has been developed. It has been implemented in the software GHQUAR.

This method which has been described in [9, 10], and [11], builds on the fact that a suppressed cell in a simple n-dimensional table without substructure cannot be disclosed exactly if that cell is contained in a pattern of suppressed, nonzero cells, forming the corner points of a hypercube.

The method subdivides n-dimensional tables with hierarchical structure into a set of n-dimensional sub-tables without substructure. These sub-tables are then protected successively in an iterative procedure that starts from the highest level. Successively for each primary suppression in the current sub-table, all possible hypercubes with this cell as one of the corner points are constructed.

For each hypercube, a lower bound is calculated for the width of the suppression interval for the primary suppression, that would result from the suppression of all corner points of the particular hypercube. To compute that bound, it is not necessary to implement the time consuming solution to the Linear Programming problem. If it turns out that the bound is sufficiently large, the hypercube becomes a feasible solution. For any of the feasible hypercubes, the loss of information associated with the suppression of its corner points is calculated. The particular hypercube that leads to minimum information loss is selected, and all its corner points are suppressed.

After all sub-tables have been protected once, the procedure is repeated in an iterative fashion. Within this procedure, when cells belonging to more than one sub-table are chosen as secondary suppressions in one of these sub-tables, in further

---

[1] European Plan for Research in Official Statistics.

processing they will be treated like sensitive cells in the other sub-tables they belong to.

It should be mentioned here that the 'hypercube criterion' is a sufficient but not a necessary criterion for a 'safe' suppression pattern. Thus, for particular subtables the 'best' suppression pattern may not be a set of hypercubes – in which case, of course, the hypercube method will miss the best solution and lead to some overprotection. So it should be noted that there is some tendency for over-suppression connected to this method. How much difference this tendency towards oversuppression makes in practice is another question.

2.1   Disclosure Control Aspects

Within this section we will explain concepts used in GHQUAR to avoid disclosure and how to use them efficiently.

Exact Disclosure.

After GHQUAR has been applied to a simple n-dimensional table (without substructure) any suppressed cell will be corner point of a hypercube where all the other corner points are suppressed as well. If all the suppressed cell values are non-zero then it is not possible to recalculate the exact value of any of the suppressions – at least if an attacker making an attempt to recalculate suppressed entries would use no other information beyond what is contained in the published table.

When the table that GHQUAR has been applied to is, however, a complex table with hierarchical substructure, then it may happen in certain cases that an attacker considering the complex structure of the table may be able of exact disclosures. This is due to the non-global method of sub-table partitioning in GHQUAR. For illustration of this problem see [10, p. 5], for general discussion of problems with non-global methods for secondary cell suppression see [3].

The technique of partitioning a complex table into separately treated sub-tables is, however, not essential for the hypercube method, and a new version of the method is in development which will use a global method to handle tables with hierarchical substructure.

Special care has been taken in the development of GHQUAR to avoid the problem of *exact insider disclosure*: A typical case of exact insider disclosure is when a single respondent cell appears as corner point in only one hypercube which does not have corner points in common with any other hypercube of the suppression pattern, but does contain, except for the single respondent cell, other sensitive cells. The single respondent, who often can be reasonably assumed to know that he is the only respondent, can use his 'insider knowledge', that is, the amount of his own contribution, to recalculate the value of any other suppressed corner point of this hypercube. To prevent this situation, or similar ones, GHQUAR makes sure that a single respondent cell will never appear to be corner point of one hypercube only, but of two hypercubes at least.

Interval Disclosure.

When the sensitive cells have been identified as such according to a concentration rule for instance, which is often the case for economic data, then it is certainly not enough for secondary suppression to prevent exact disclosure only. The idea behind a concentration rule is to prevent that users of statistical data are able to obtain close estimates for individual respondent data. In choosing the particular parameters of the concentration rule the disseminator defines the precision of an (upper) estimate that would be too close. Considering this, the secondary suppressions must also be selected in such a way that users of a table published with suppressions are not able to derive estimates for individual respondent data that are any closer.

It is a well known fact that it is possible to derive upper and lower bounds for the true value of any particular suppressed entry of a table with non-negative, or similarly bounded, entries by using the linear relations between published and suppressed cell values in the table, and eventually some additional *a priori* constraints on the cell values. The interval given by these bounds is called the 'suppression interval'.

The upper bound of the suppression interval can be used to derive upper estimates for individual contributions to the suppressed cell. It has been proven ([2,3]; for illustration and example see [7]) that these upper estimates will be too close according to the concentration rule of the disseminator, if the upper bound of the suppression interval for a sensitive cell is below a certain minimum feasible size which depends on the particular distribution of the individual contributions to this cell. (See [5], Appendix) for formulas for minimum feasible upper bounds for the most common sensitivity rules.) Note that, in relation to the cell total, the feasible upper bound will be the larger, the stronger the concentration of the contributions in the cell, with a (relative) maximum for the worst case which are the single respondent cells. This fact can be used to derive (worst case) feasible upper bounds, that are unrelated to the particular distribution of the individual contributions in a cell, but depend on the cell value only. Such bounds can then be stated as a ratio to the cell total. See the appendix of this paper for formulas relating to the most common sensitivity measures.

Users of GHQUAR are requested to state a minimum width for the suppression interval, that is, the size of a *sliding protection range,* as a ratio to the cell total. When choosing a hypercube to protect a particular sensitive cell, GHQUAR computes the width of the suppression interval that would result from suppressing all its corner points, while eventually considering given *a priori* constraints for the cell values, and assuming that there are no other suppressions within this (sub-) table. (Note, that under this assumption the width of the suppression interval is identical for all corner points.) The hypercube is considered to be feasible only if the computed width exceeds the size of the sliding protection range as stated by the user.

In the following we propose two alternative options to control GHQUAR as to make sure that the upper bound of the suppression interval for any sensitive cell exceed the minimum feasible size mentioned above:

1. Set the size of the sliding protection range to the ratio r given in the appendix according to the sensitivity measure used for primary suppression, or instead
2. Set fixed q percentage upper and lower *a priori* constraints for any cell (for illustration see 2.3 below), and set the size of the sliding protection range to

$2\frac{q}{100}(r-1)$, where r equals the ratio given in the appendix. Note, that the percentage q corresponds to the second parameter of a (p,q)-rule and should be set to 100 in case $p\%$- or (n,k)-dominance rules have been used for primary suppression.

Experience with the first option which is based on the implicit assumption of zero lower bounds of the protection interval has shown a strong tendency towards over-suppression, while the second alternative builds on a very recent suggestion, the performance of which has not yet been tested in practice.

Regarding the problem of *interval insider disclosure*, such as, for instance, the situation that a contributor to a sensitive cell can closely estimate the contribution of one of the contributors to another sensitive cell in the same row (or column, etc.), a satisfying solution how to make the program avoid such a constellation in the suppression pattern has not yet been found. A method to solve the problem, however, has recently been suggested, and may prove to be feasible.

2.2 Information Loss Aspects

For any hypercube satisfying the disclosure control conditions as described in Section 2.1, GHQUAR computes the loss of information associated with the suppression of its corner points. The particular hypercube that leads to minimum information loss is selected, and all its corner points are suppressed.

The standard information loss measure for this selection procedure (in the following also referred to as 'costs' for suppressing a cell) is proportional to the logarithm of the cell value. This measure is used in combination with some heuristic approaches to distinct between several categories of cells, such as suppressed and (so far) unsuppressed cells, zero cells, and single contributor cells. The costs for suppressing a hypercube that contains at least one unsuppressed cell, for instance, will at any rate exceed the costs for a hypercube containing only cells that are already suppressed.

There is also an option to make the algorithm avoid the suppression of cells in the margins of subtables. With this option, cells in the margins of a subtable are assigned high additional costs. The use of this option is strongly recommended – not just because the information content of cells in margins may be higher, but because these cells create links between the subtables of a complex table. Any suppressed cell in a margin of one subtable may produce additional suppressions in another subtable.

Another option supports a certain kind of 'local recoding'. With this option, used for instance, when one of the table dimensions is a size classification, adjacent cells (such as for instance a cell in the adjacent size class) are preferred as complementary suppressions. This allows in certain cases to 'recode' the size classification locally, that is, to publish the value of the response variable for the combination of those two adjacent size classes for a particular combination of the variables defining the other dimensions of the table. The implementation of this heuristic is such that the costs for suppressing a cell a weighted. The weight depends on how close cells are to a target suppression. Direct 'neighbors' are assigned low weights.

Apart from this specific weighting functionality which is automated (the user only 'switches on' this option for a particular dimension of the table, and then the program will generate suitable weights automatically), there is also an option for 'manual' weighting. With this 'manual' weighting option, the user assigns weight to particular cells directly, which may affect the likelihood for those cells to be selected as complementary suppressions. It should, however, be noted that this option is in conflict with the option to avoid cells in the margins of tables to be suppressed – it is not allowed to use both options at the same time. Because, as explained above, it is usually important to avoid cells in the margins to be suppressed in order to prevent a certain kind of oversuppression, a support routine is offered (see 4.2 below) providing for automated generation of suitable weights which will not cause this conflict.

2.3   Control Options

Within this section we will illustrate some more control facilities of GHQUAR, in addition to those mentioned in 2.1 and 2.2 . We will, however, concentrate here on those control options, the decision of whether or not and in which way to use them, will probably still be up to the user, even if access to GHQUAR is through a control center such as τ-ARGUS.

- *A priori* constraints: If it must be assumed that in advance of publishing a table the cells of the table are approximately known, e.g. that potential users of the table can estimate cells to within certain upper and/or lower bounds, then this information can be passed to GHQUAR, which will in that case choose suppression patterns in such a way that this knowledge will not cause disclosure. *A priori* constraints are specified to GHQUAR as deviations from the original cell values. As mentioned in 2.1, use of *a priori* constraints may turn out to be useful for the selection of complementary suppressions even when the disseminator does not assume such approximate knowledge to be around. To specify, for instance, fixed q percentage *a priori* bounds as recommended in 2.2, set both, lower and upper deviations to q % of the cell value.
- Tables containing both, positive and negative values: GHQUAR may process those tables as well. The user of GHQUAR may choose whether to replace negative values by their absolutes, or add a large constant, thus shifting all negatives to positives. In taking this decision, the user should of course take into account the effect that his choice may have considering the GHQUAR interval disclosure control and information loss methods described above.
- Eligibility of zero cells for secondary suppression: When, at least for some zero cells, their location is not assumed to be known in advance of publishing a table, and the disseminator considers it sufficient to prevent *exact* disclosure of the data, then, in certain cases, zero cells may serve as secondary suppressions. Those cells of *a priori* unknown location must be assigned a large positive value in the data set (exceeding the largest of the original values) to make them eligible for secondary suppression.

3   Applying GHQUAR to Overlapping Tables: The GHMITER Algorithm for Table-to-Table Protection

Usually some of the tables in the set of multiple tables published from the same source (*e.g.*, response data from a survey) will be overlapping. Let for instance, a table T1 present 'turnover by enterprise employee size class', a Table T1.1 present 'turnover by NACE and enterprise employee size class', and a Table T1.2 present 'turnover by enterprise employee size class and enterprise legal form'. Then T1 is a subtable of T1.1, as well as of T1.2 , at least if all the categories of 'employee size class' are identical for both tables T1.1 and T1.2 . A cell of the overlap-table T1 will be a sensitive cell of T1.1 if, and only if, this cell is also a sensitive cell of T1.2 .

When secondary cell suppression is carried out for T1.1 and T1.2 individually, it is not unlikely that there will be T1 cells unsuppressed in T1.1 that are complementary suppressions in T1.2, and *vice versa*. Any user given access to both tables T1.1 and T1.2 will be able to disclose these values, and may hence be able to recalculate sensitive cells.

There are ways of preventing this situation. One alternative would be to protect the 'full' table T1.3: 'turnover by enterprise employee size class, NACE, and enterprise legal form' and suppress in T1.1 and T1.2 any cells which also were suppressed in T1.3 . Assume now, maybe due to an exceedingly fine employee size class scheme, it is not possible to protect this table within a single run, because of huge computer resource requirement. In this situation, a table-to-table protection procedure is needed.

Such a procedure has recently been provided by the developers of GHQUAR. Technically, the table-to-table protection tool GHMITER for protection of overlapping tables was implemented by taking the individual subroutines of GHQUAR and putting them (with some modification) together in a new way[2]. The result is a software that can be applied to protect a set of multiple, overlapping, complex tables in iterative fashion. Each table is protected separately, while the algorithm keeps track of any new suppressions belonging to overlap sections. The procedure is repeated until no cells get newly suppressed in any of the overlap sections.

In the example above, GHMITER will first apply secondary cell suppression to, for example, Table t1.1, and then to table T1.2, keeping track of any secondary suppressions in the overlap table T1. Secondary suppressions in T1, as resulting from protecting T1.1 will be treated like primary suppressions when protecting T1.2, and *vice versa*. The procedure will be repeated over and over again, until a step of the iteration is reached where no new secondary suppressions have been selected in T1. After the table-to-table protection procedure is finished, any cell of the overlap table T1 is either suppressed in both T1.1 and T1.2, or unsuppressed in both T1.1 and T1.2 . Moreover, none of the suppressions can be disclosed by making use of the additive relationship between suppressed and unsuppressed cells in either T1.1 or T1.2 .

---

[2] In fact, the 'pure' GHQUAR software is not supported any more. Users are requested to use GHMITER also for protecting single tables.

4   Co-ordination of Suppression Patterns: The Support Routine POOLAC

Ideally, a table-to-table protection procedure should be applied to the full set of tables potentially releasable from a data source. This option seems less and less realistic, however, as technological advances progressively change the process of statistical data production and release. Formerly, the set of cells/tables published from a particular survey would have been largely fixed in advance. Now, however, the process of releasing data is becoming more and more user demand driven and less preplanned, even to the extent of providing public use statistical data base query systems. Inability to forward-plan causes serious trouble for cell suppression.

Moreover, problems arise when data are published on different levels of a regional classification (e.g. on the national and on the super national (EU) level, or on the regional and national level) but secondary suppressions are to be assigned by different agencies actually (e.g. national statistical institutes and Eurostat, or regional and national statistical institutes).

Need for co-ordination of suppression patterns arises also in the situation of a table published periodically, monthly, quarterly, or annually for instance. A part of the sensitive cells may be sensitive in every period. As simple illustration, assume a table without substructure, containing some sensitive cells, which are 'for ever' sensitive. Assume further, that there is more than one feasible suppression pattern, and that the costs for each pattern differ only slightly. If nothing is done, it is then very likely that the suppression pattern changes from period to period, which might be undesirable, and also cause a risk of disclosure, when the variation in the cell values of the secondary suppressions for different periods is only small.

It is one of the tasks of the CASC project to research heuristic strategies to solve these problems using facilities such as provided by the support routine POOLAC which we will describe in Sections 4.1 and 4.2 below.

4.1   Facing the Problem of User Demand Driven Table Production

The situation can be improved to some extent, when data are 'pooled' to keep track of all suppressions in tables already released. One might even attempt to use such a data pool as data basis for public- or scientific use data base query systems.

The software POOLAC, implemented at the German Federal Statistical Office, is able of setting up such a 'data pool'. The data pool will contain one and only one record for each cell of any table already protected and this record contains an entry regarding the suppression status of the cell. In advance of protecting a new table (with GHMITER, for instance) POOLAC will investigate the data pool for any cell of the new table. If any cell has already been used as secondary suppression in one of the tables processed earlier, then, in the POOLAC output file, which is supposed to be used as input file for GHMITER, this cell will be flagged suppressed. If, on the contrary, the suppression status for the cell is 'unsuppressed' according to the data pool entry, POOLAC will assign a suitable preference code (see Section 4.2 below) to this cell, which will make GHMITER avoid to select it as complementary suppression in the new table To some extent.

4.2   Specialized Facilities to Support the Co-ordination of Suppression Patterns

Such facilities, which we call 'preference facilities' are supposed to make GHMITER prefer or even force certain cells to remain unsuppressed, or on the contrary, to make certain cells be used as complements first.

The program POOLAC provides options to define arbitrary subsets of cells of a table - in the extreme a set may even contain a single cell only. The user may then assign a 'preference code' to such a set of cells. POOLAC will enter this preference code into those records of the output data file that relate to any of the cells in the set.

Use of this output data file as input to GHMITER will make GHMITER increase, or, on the contrary, decrease the costs for suppressing a cell, depending on the particular value of the preference code. In the current prototype implementation, the preference codes correspond each to one of six different algorithms to compute increased or decreased costs. There is further a preference code offered that causes GHMITER to assume *a priori* constraints making cells ineligible for suppression. At the stage of writing this, however, no practical experience on the performance of these methods has yet been gained.

5   Graphic User Interfaces for GHQUAR

To be able to run GHQUAR (or GHMITER), apart from providing control information, the user must also supply information on the structure of the table To be protected. The programs require sorted data files, code lists in a certain format, and, moreover, the codes of the key-variables must be provided in a certain format. This makes the data preparation process somewhat uncomfortable and also error-prone, to the extent even that technology transfer of the programs to agencies outside of Germany seems unrealistic. What is needed is a Graphic User Interface, to assist and guide users in providing the required input information. Sections 5.1 and 5.2 illustrate plans and recent developments supposed to improve access to the programs.

5.1   Improving Accessibility of GHQUAR with τ-ARGUS

When accessing the programs through the first version of τ-ARGUS to support GHMITER, data preparation will be much easier for the user. All that must be done is to supply the data-set with hierarchical codes (any kind of hierarchical codes will do). Then τ-ARGUS will be able to extract the structural information from the data-file. It will rewrite the data-file according to the needs of GHMITER and will also prepare the control information in the required formats.

Later versions should go still beyond that, encouraging the user to 'play' with the data, to, for instance, redesign the table over and over again, drop sub-totals or introduce new ones, and so forth, until finally he ends up with a favorable table and an acceptable suppression pattern. Later versions of τ-ARGUS will particularly support the functionality of POOLAC. There might be options provided to define 'preferences' in automated fashion, like a switch to be put on or off for certain classes of cells, such as e.g. sub-total cells, particular cells of overlapping parts of tables, cells used as secondary suppressions in a previous period for tables presenting results of periodical surveys, etc.

In such a completed version, τ-ARGUS should also set up meta files with historical information on the data pool, showing which tables are already contained in the pool, with information on the run-parameters used and log-files created when protecting them. It will be simple to extract data from the pool file into ready-made tables.

5.2   Enhancing First User Experience: The Eurostat GUI

Prior to the CASC project, Eurostat has decided to make an attempt to use GHQUAR for protection of European level data. To ease this first user experience, on behalf of Eurostat a first graphic user interface CIF (Confidentiality Interface, documentation in [1]) has been developed, and is currently being updated to interface with GHMITER.

Tabular data protection is a very complex business. Users of software for tabular data protection must have a good understanding of the methods used to select secondary suppressions. In order to come to a good understanding of those methods it is certainly helpful to gain some practical experience with their application. On the other hand, it is essential in software development to receive qualified user feed-back. The Eurostat CIF will enlarge the GHQUAR/GHMITER user community and may thus lead to valuable user feed-back which may also turn out to be of use in the development of τ-ARGUS. It can thus be expected, that development of the CIF will have a sustaining effect on work to be done in the CASC project.

6   Final Remarks

It is a general objective of the EU project CASC to integrate best praxis tools and methods for statistical confidentiality into the ARGUS software. One of those tools is the GHQUAR algorithm for secondary cell suppression [6].

It has been the main issue of this paper to provide the kind of basic knowledge on GHQUAR as will have to be transferred to users who gain access to GHQUAR through ARGUS or a similar interface software. Apart from this, the paper is supposed to serve as a basis for internal discussion and knowledge transfer within the CASC team in the process of building the link between ARGUS and GHQUAR.

The paper has illustrated the control options of GHQUAR, in particular how to control GHQUAR to avoid certain disclosure risks on the enterprise level (Section 2.2), while also mentioning certain residual disclosure risks that cannot yet be avoided with this method. It could be mentioned, however, that this kind of residual disclosure can hardly be fully avoided by any other current method. The paper has presented the concept of GHQUAR, or the related program GHMITER respectively, for table-to-table protection of overlapping tables and even more general concepts to co-ordinate suppression patterns between tables.

Acknowledgements

This work was partially supported by the EU projects IST-2000-25069 Computational Aspects of Statistical Confidentiality, and IST-2000-26125 Accompanying Measure to R&D in Statistics

References

1. Biermann, V., Förster, J., and Planes, F. (2000): 'User Manual of the Confidentiality_Interface (CIF v1)', unpublished Manual, Eurostat, Luxemburg, July 2000
2. Cox, L. (1981), 'Linear Sensitivity Measures in Statistical Disclosure Control', Journal of Planning and Inference, 5, 153 - 164, 1981
3. Cox, L. (2001), 'Disclosure Risk for Tabular Economic Data', In: *'Confidentiality, Disclosure, and Data Access: Theory and Practical Applications for Statistical Agencies'* Doyle, Lane, Theeuwes, Zayatz (Eds.), North-Holland
4. Gießing, S. (1998), 'Looking for efficient automated secondary cell suppression systems: A software comparison', Research in Official Statistics Journal 2/98
5. Gießing, S. (2001), 'New tools for cell suppression in τ-ARGUS: One piece of the CASC project work draft', paper presented at the Joint ECE/Eurostat Worksession on Statistical Confidentiality in Skopie (The former Yugoslav Republic of Macedonia), 14-16 March 2001
6. Giessing, S. (2001), 'The CASC Project: Integrating Best Practice Methods for Statistical Confidentiality', in pre-proceedings of the NTTS & ETK 2001 Conference, Hersonissos (Crete) 18-22 June 2001.
7. Giessing, S. (2001), 'Nonperturbative Disclosure Control Methods for Tabular Data', In: *'Confidentiality, Disclosure, and Data Access: Theory and Practical Applications for Statistical Agencies'* Doyle, Lane, Theeuwes, Zayatz (Eds), North-Holland
8. Hundepool, A. (2001), 'The CASC project', paper presented at the Joint ECE/Eurostat Worksession on Statistical Confidentiality in Skopie (The former Yugoslav Republic of Macedonia), 14-16 March 2001
9. Repsilber, R. D. (1994), 'Preservation of Confidentiality in Aggregated Data', paper presented at the Second International Seminar on Statistical Confidentiality, Luxemburg, 1994
10. Repsilber, D. (1999), 'Das Quaderverfahren' - in *Forum der Bundesstatistik, Band 31/1999: Methoden zur Sicherung der Statistischen Geheimhaltung*, (in German)
11. Repsilber, D. (2000), 'Wahrung der Geheimhaltung sensibler Daten in mehrdimensionalen Tabellen mit dem Quaderverfahren' - in *Statistische Analysen und Studien Nordrhein-Westfalen, Landesamt für Datenverarbeitung und Statistik NRW, Ausgabe 3/2000* (in German)

Appendix

Minimum Feasible Upper Bounds for the Protection Interval

Sensitivity rule	Ratio to the cell total for worst case minimum feasible upper bounds for the protection interval for a sensitive cell
(n,k)-rule	$\dfrac{100}{k}$
p%-rule	$\dfrac{p+100}{100}$
(p,q)-rule	$\dfrac{p+q}{q}$

SDC in the 2000 U.S. Decennial Census*

Laura Zayatz

Bureau of the Census, Statistical Research Division
3209-4, Washington, D.C. 20233

Abstract. This paper describes the statistical disclosure limitation techniques to be used for all U.S. Census 2000 data products. It includes procedures for short form tables, long form tables, public use microdata files, and an online query system for tables. Procedures include data swapping, rounding, noise addition, collapsing categories, and applying thresholds. Procedures for the short and long form tables are improvements on what was used for the 1990 decennial census. Several procedures we will be using for the microdata are new and will result in less detail than was published from the 1990 decennial census. Because we did not previously have the online query system for tables, all of those procedures are newly developed.

1 Introduction

The Bureau of the Census is required by law (Title 13 of the U.S.Code) to protect the confidentiality of the respondents to our surveys and censuses. At the same time, we want to maximize the amount of useful statistical information that we provide to all types of data users. We are in the process of applying the disclosure limitation techniques for all data products stemming from Census 2000. The techniques are designed to protect data confidentiality while preserving data quality.

This paper describes the Census 2000 disclosure limitation techniques. In Section 2, we briefly describe the procedures that were used for the 1990 Census. In Section 3, we explain why some changes in those techniques were necessary. In Sections 4-7, we describe the procedures for Census 2000, including procedures for the 100% (short form) census tabular data, the sample (long form) tabular data, the microdata, and the online query system (Tier 3) in American FactFinder. Section 8 offers a conclusion.

---

*This paper reports the results of research and analysis undertaken by Census Bureau staff. It has undergone a Census Bureau review more limited in scope than that given to official Census Bureau publications. This report is released to inform interested parties of ongoing research and to encourage discussion of work in progress.

J. Domingo-Ferrer (Ed.): Inference Control in Statistical Databases, LNCS 2316, pp. 183 - 202, 2002.
© Springer-Verlag Berlin Heidelberg 2002

2   Disclosure Limitation for the 1990 Census

2.1   Procedure for the 100% (Short Form) Data

The Census Bureau attempts to get information on characteristics such as sex, age, Hispanic/NonHispanic, race, relationship to householder, and tenure (owner or renter) from 100% of the population through what we call "short form" questions. The 100% data are published in the form of tables. Many of the tables are published at the block level. The average block contains 34 people. Some of the more detailed tables are published at the block group level. The average block group contains 1,348 people. Thus these data are published for very small geographical units. For the 1990 census, the procedure used to protect the short form (100%) data was the Confidentiality Edit [1]. A small sample of census households from the internal census data files was selected. The data from these households were swapped with data from other households that had identical characteristics on a certain set of key variables but were from different geographic locations. Which households were swapped was not public information. The key variables were number of people in the household of each race by Hispanic/NonHispanic by age group (<18,18+), number of units in building, rent/value of home, and tenure. All tables were produced from this altered file. Thus census counts for total number of people, totals by race by Hispanic/NonHispanic by age 18 and above (Public Law 94-171 counts --- also known as Voting Rights counts) as well as housing counts by tenure were not affected. A higher percentage of records was swapped in small blocks because those records possess a higher disclosure risk. All data from the chosen households were swapped except for Indian Tribe. It was felt that it did not make sense to move a member of one tribe into a location inhabited by another tribe.

One advantage of the Confidentiality Edit is that it only needs to be implemented once on the internal microdata file in order to protect all tables produced from the file. A requirement for Tier 3 of American FactFinder, a table query system described in Section 3.3, is that the majority of disclosure limitation techniques be applied to the underlying data rather than to individual tables. We wish to avoid techniques such as random rounding, cell suppression, and perturbation which are often applied on a table by table basis. An additional advantage of the confidentiality edit is that no data are suppressed, so aggregation of data is not a problem. The disadvantage is that there are no obvious changes in the tables that would make evident our disclosure limitation efforts.

2.2   Procedures for the Sample (Long Form) Data

Sample Data in Tabular Form.
Approximately a one in six sample of the population receives the long census form which, in addition to the 100% information, collects information on characteristics such as marital status, school attendance and grade level, ancestry, language, place of birth, citizenship, military service, income, industry, and occupation. The sample data are also published in the form of tables. Some of the tables are published at the block group level.

The average block group contains 1,348 people. Some of the more detailed tables are published at the tract level. The average tract contains 4,300 people. Thus these data are also published for very small geographical units. In 1990, it was felt that the fact that it was a sample provided protection for all areas for which sample data were published except for small block groups. In small block groups, some values from one housing unit's record on the internal file were blanked and imputed using the 1990 census imputation methodology. This altered file was used to create all tables. Which values were altered was not public information.

Sample Data in Microdata Form.
We collect sample data from approximately 17% of all households. From the 1990 census, we published a 5% state file, a 1% metropolitan area file, and a 3% elderly file. All three Public Use Microdata Samples (PUMS) were mutually exclusive and the public use microdata areas (PUMAs) identified on all 3 files had at least 100,000 persons in them. The 5% file contained data for 5% of all households in the country, and the PUMAs followed state boundaries. The 1% file contained data for 1% of all households, and the PUMAS were allowed to cross state boundaries. The 3% elderly file contained data for 3% percent of all households with at least one person of age 60 or older. This file was requested and paid for by the Administration on Aging (AOA).

The microdata files were created from the internal file after the blanking and imputation described above, so some protection was provided by that procedure. Income values and some other continuous values such as age and rent were topcoded. Some very detailed categories from items such as Ethnicity and Indian tribe were collapsed into broader categories. And, of course, all identifying information such as name and address were stripped from the file.

3   Why Should the 1990 Procedures Be Changed?

3.1   Main Improvement: Targeting the Most "Risky" Records

As we stated in Section 2, small blocks and small block groups had higher rates of swapping and blanking and imputation because the records from small geographic areas possess a high disclosure risk. We would like to extend the idea of targeting the most risky records for the disclosure limitation techniques. We would only swap records that were unique based on some set of key variables. Those are the records with the most risk. We would not swap households for which all data were imputed. They are not at risk. We would take into account the protection already provided by the rate of imputation. Records representing households containing members of a race category, which appears in no other household in that block, are easily identifiable and present a special risk. A very large percent of those records will be swapped. And finally, we would let the swapping rate differ among blocks and have an inverse relationship with block size (in terms of number of households). We believe it would be easier to identify a person or household in a small block than it would be in a large block.

3.2   Multiple Race Issues

In 1990, a person could only be identified by a single race. That is, people were only supposed to check one box on the questionnaire in response to the race question. In 2000, people were asked to check more than one box, if applicable. Thus we now have 63 possible answers to the race question. This leads to changes in disclosure risk as well as processing procedures because of the additional detail in the tables and microdata.

3.3   American FactFinder (AFF)

AFF [2] has been developed to allow for broader and easier access to the data and to allow users to create their own data products. The goal of Tier 3 of AFF is to allow users to submit requests for user-defined tabular data electronically. A request would pass through a firewall to an internal Census Bureau server with a previously swapped, recoded, and topcoded microdata file. The table would be created and electronically reviewed for disclosure problems. If it was judged to have none, the table would be sent back electronically. This is a new way of publishing tabular data, so we needed to develop new disclosure limitation procedures for AFF.

4   The Procedure for the 100% Data

As we did in 1990, we swapped a set of selected records. Unlike 1990, the selection process was highly targeted to affect the records with the most disclosure risk. There was a threshold value for not swapping in blocks with a high imputation rate. Only records which were unique in their block based on a set of key demographic variables were swapped. A unique record was selected for swapping with a probability of:

$$C_1 * \frac{1}{block\ size} \ .$$

That is, the probability of being swapped had an inverse relationship with block size. In addition, records representing households containing members of a race category which appeared in no other household in that block had an additional P_1 probability of selection. Pairs of households that were swapped matched on a minimal set of demographic variables. All data products are created from the swapped file.

5   The Procedure for the Sample Data in Tabular Form

Swapping (rather than blanking and imputation) will be performed to protect the data. This will increase the amount of distortion (giving us more protection). Swapping has the nice quality of removing any 100% assurance that a given record belongs to a given household. It is consistent with the 100% procedure. And, it retains relationships among the variables for each household.

As with the 100% data, we will use 2 different sets of key variables - one to identify the unique records and one to find the swapping partners. We will hold several variables fixed (unswapped). For example, travel time to work and place of work for a household may not make sense if swapped with a household geographically far away.

The procedure for producing the masked file then is very similar to the procedure for the 100% data. Blockgroup replaces block because blockgroup is the lowest level of geography for publishing sample data. The threshold value for not swapping in blockgroups with a high imputation rate may differ, and the probability of a unique record being swapped is:

$$C_2 * \frac{sampling\ rate\ for\ that\ blockgroup}{blockgroup\ size}.$$

We have given the chance of being swapped an inverse relationship with blockgroup size. We have also given the chance of being swapped a direct relationship with blockgroup sampling rate. The lower the sampling rate, the more likely that the sample unique is not unique in the entire blockgroup population. So a smaller sampling rate should lead to a lower chance of being swapped.

6   The Procedure for the Sample Data in Microdata Form

An internal examination of the microdata in conjunction with the tables published from the 1990 census and an increased concern about external data linking capabilities prompted the Census Bureau to propose a decrease in the detail provided on the 2000 PUMS. There will be a 5% state file where PUMAs must contain at least 100,000 persons and follow state boundaries. This file will contain less detail for several variables than in 1990. There will be a 1% characteristics file with "super- PUMAs" which must contain at least 400,000 persons and follow state boundaries. This file will contain approximately the same level of detail for most variables as in 1990. The files will be mutually exclusive. The 1% characteristics file will be released in 2002. The 5% state file, requiring more time for post-processing, will be released in 2003.

There will be some changes in detail from 1990 for both the 5% state file and the 1% characteristics file. Previously, dollar amounts (wages and salary, rent, etc.) were published to the exact dollar. For Census 2000, all dollar amounts will be rounded according to the following scheme:

$1-7=$4
$8-$999 round to nearest $10
$1,000-$49,000 round to nearest $100
$50,000+ round to nearest $1,000

The rounding will be done prior to all summaries and ratio calculations.

Previously, departure time to work was published to the exact minute. For Census 2000, departure time will be rounded as follows:

2400-0259 in 30-minute intervals
0300-0459 in 10-minute intervals
0500-1059 in 5-minute intervals
1100-2359 in 10-minute intervals

Previously, noise was not added to any variables on the PUMS. For Census 2000, noise will be added to the age variable for persons in households with 10 or more people. Ages will be required to stay within certain groupings so program statistics are not affected. Original ages will be blanked, and new ages will be chosen from a given distribution of ages within their particular grouping.

There will be some changes affecting only the 5% state file. For the 5% state file, all categorical variables must have at least 10,000 persons nationwide in each published category. This reduces much of the detail in some of the variables in this file as seen below.

Table 1. Number of Categories for Selected PUMS Variables in the 5% State File

	1990	2000
Language	305	74
Ancestry	292	143
Birthplace	312	167
States	51	51
Puerto Rico and Island Areas	8	5
Foreign Born	253	111
American Indian Tribe	27	23
Hispanic Origin	48	29
Marital Status	6	6
Occupation	506	443
Industry	245	245

These numbers are based on 1990 census data. The actual Census 2000 categories will be identified through a post-processing of the data. Because of the increase in the population size over the last decade, there may be a few more categories that meet the threshold. Users will be consulted as to how best to combine categories that do not meet the 10,000 person threshold.

Race is the one variable where users will receive more data in the 2000 PUMS than in the 1990 PUMS. In 1990, a person could only be identified by a single race. That is, people were only supposed to check one box on the questionnaire in response to the race question. In 2000, people were asked to check more than one box, if applicable. The fifteen eligible boxes were White, Black, American Indian or Alaska Native, Asian Indian, Chinese, Filipino, Japanese, Korean, Vietnamese, Other Asian, Native Hawaiian, Guamanian or Chamorro, Samoan, Other Pacific Islander, and Some Other Race. Note that if a person checked American Indian or Alaska Native, they were asked to write in the name of the enrolled or principal tribe, and if they checked Other Asian, Other Pacific Islander, or Some Other Race, they were asked to write in the name of the race. Also note that Asian Indian, Chinese, Filipino, Japanese, Korean, Vietnamese, and Other Asian are subgroups of Asian, and Native Hawaiian, Guamanian or Chamorro, Samoan, and Other Pacific Islander are subgroups of Native Hawaiian or Pacific Islander. Thus the six major race groupings are White, Black, American Indian or Alaska Native, Asian, Native Hawaiian or Pacific Islander, and Some Other Race.

On the 2000 PUMS, there will be a YES/NO variable for each of the six major race groupings. So, for example, a user can see if the respondent was both Black and Asian. There will also be a variable similar to one that was on the 1990 PUMS which allows for the designation of one of approximately 70 single race groups including some detailed groups of American Indian tribes, detailed Asian groups, and detailed Native Hawaiian or Pacific Islander groups that were write-ins. On the 5% file, each of these groups must meet the 10,000 person threshold, and on the 1% file, each of these groups must meet an 8,000 person threshold. In addition, there will be a new variable which allows for the designation of all combinations of 2 or more races. For this variable, a 10,000 person threshold applies to the 5% file, and an 8,000 person threshold applies to the 1% file. So, for example, a user can see if the respondent was both White and Chinese.

For both the 5% file and the 1% file, after the PUMS sample has been chosen, there will be a second round of data swapping, also done at the household level as described in section 5. We will examine the records, looking for what are often called "special uniques." These are household records which remain unique based on certain demographic variables at very high levels of geography and therefore have a high level of disclosure risk. Any such household we find will be swapped with some other household in a different PUMA or SuperPUMA. We do not believe that this will affect many records, but those that it does need this added protection.

There are some procedures that have not changed. Topcoding is used to reduce the risk of identification by means of outliers in continuous variables (for example someone with an income of five million dollars). As in 1990, all continuous variables (age, income amounts, travel time to work, etc.) will be topcoded using the half-percent/three-percent rule. Topcodes for variables that apply to the total universe (for example age) should include at least 1/2 of 1 percent of all cases. For variables that apply to subpopulations (for example farm income), topcodes should include either 3 percent of the non-zero cases or 1/2 of 1 percent of all cases, whichever is the higher topcode. Some variables such as year born will be likewise bottomcoded.

And as in 1990, because the variable Property Taxes is readily, publicly available, it will be put into larger categories than those resulting from the rounding described above.

7   The Procedures for American FactFinder

American FactFinder does not provide an open-ended or unconstrained opportunity to construct any or all possible tabulations from the full microdata files. As stated previously a query for a table through AFF would pass through a firewall to an internal Census Bureau server with a previously swapped, recoded, and topcoded microdata file. All tables generated from the sample data will be weighted. The query and the resulting table must each pass through a filter.

7.1   The Query Filter

If a user requests a tabulation for more than one area or for a combination of areas, each area must individually pass the query filter. Guidelines for requesting tabulations will include:

o   Levels of geography: The external user is advised in the user interface that the blockgroup is the lowest level of geography permitted for 100% data and the tract is the lowest level of geography permitted for sample data for an external user. Requests for split blockgroups or split tracts are not permitted. A minimum population requirement is also imposed.

o   Maximum number of table dimensions: The user interface permits no more than 3 dimensions (page, column, and row) not including geography.

o   Total population per geographic unit: Population size criteria will be determined from data in the summary files that indicate whether the population is large enough to pass the results filter. The user will be informed if the population size is too small.

The query filter also delimits the use of sensitive variables such as race, Hispanic origin, group quarters, cost of electricity, gas, water, fuel, property taxes, property insurance cost, mortgage payments, condo fees/mobile home costs, gross rent, selected monthly owner cost, household/family income and individual income types. External users may obtain only predefined categories or recoded values of these variables.

The system determines if the query requests small areas, that is less than average tract size (4300), medium areas (population size 4301-99,999) or large areas (population size 100,000 or more). According to the population size of the area or areas requested, the system permits the use of appropriate combinations of short (least detailed), medium, or long (most detailed) lists of predefined categories of race, Hispanic origin, group quarters and other sample variables in the cross-tabulation. Only topcoded values of sensitive variables may be accessed.

If the query passes the query filter rules, the query is sent from the external server outside the firewall to the internal server inside the firewall to the full microdata files. The full microdata files contain all of the predefined categories for race, Hispanic origin, group quarters and modified sensitive variables.

7.2   The Results Filter

Each resulting tabulation selected from the full microdata files obtained through American FactFinder must meet certain criteria or American FactFinder will not provide the user with the tabulation. If a user requests a tabulation for more than one area or for a combination of areas, each area must individually pass the results filter. The criteria are designed to prevent the release of sparse tabulations which can lead to disclosure. If a tabulation does not meet the criteria, the user will receive a message stating that the tabulation cannot be released for confidentiality reasons.

The system computes the total mean and median population cell sizes of the tabulation. For both mean and median calculations, only the internal cell counts are used (not the marginal totals). For both the mean and median calculations, cells with zero are included. If either the mean or median is less than some number n the system does not permit the tabulation.

Our disclosure limitation rules are designed to prevent the release of sparse tables. They do not guarantee that there will be no cell values of 1. In fact, many of our predefined tables contain cell values of 1, and for those we rely on the data swapping procedure to protect the data. American FactFinder uses the swapped file in generating tables, but again we wish to avoid releasing very sparse tables. The third rule in the results filter limits the proportion of cells with values of one. The rule counts the total number of nonzero cells in the cross-tabulation and the number of cells in the cross-tabulation with a value of 1 and then ensures that the ratio of the count of cells with a value of 1 to the count of total nonzero cells in the cross-tabulation is less than some preset parameter.

8   Conclusion

Prior to production, we tested and evaluated the data swapping procedures with various parameters using data from the 1995 and 1996 Census tests and the 1998 Dress

Rehearsal. We are in the process of evaluating the procedure's performance on the Census 2000 data in terms of reducing disclosure risk and maintaining data quality.

Working very closely with data users, the Census Bureau was able to develop specifications for the PUMS which will protect the confidentiality of the data and provide the maximum quality and quantity of data to the users.

The American FactFinder rules and their parameters and population threshold values were tested by Census Bureau staff and disclosure limitation experts at Carnegie Mellon University. They were judged to be in accordance with best practices.

References

1.   Griffin, R., Navarro, F., and Flores-Baez, L. (1989), "Disclosure Avoidance for the 1990 Census," *Proceedings of the Section on Survey Research Methods*, American Statistical Association, pp. 516-521.
2.   Rowland, S. and Zayatz, L. (2001), "Automating Access with Confidentiality Protection: The American FactFinder," *Proceedings of the Section on Government Statistics*, American Statistical Association, to appear.

Applications of Statistical Disclosure Control at Statistics Netherlands

Eric Schulte Nordholt

Statistics Netherlands, Division Social and Spatial Statistics
Department Support and Development, Section Research and Development
P.O.Box 4000, 2270 JM Voorburg, The Netherlands
esle@cbs.nl

Abstract. Results of European research projects on SDC (Statistical Disclosure Control) are used in the production of official statistics in the Netherlands. Two related software packages are described that can be applied for producing safe data. The package τ-ARGUS is used for tabular data and its twin μ-ARGUS for microdata. Both τ-ARGUS and its twin μ-ARGUS are products developed in the SDC project under the Fourth Framework Programme of the European Union. New versions (that include results of the on-going research) of both packages will be released in the CASC (Computational Aspects of Statistical Confidentiality) project that is funded under the Fifth Framework Programme of the European Union. Bona fide researchers who need more information have the possibility to visit Statistics Netherlands and work on-site in a secure area within Statistics Netherlands. Some examples are given of official statistics in the Netherlands that have benefited from statistical disclosure control techniques.

1   Introduction

The task of statistical offices is to produce and publish statistical information about society. The data collected are ultimately released in a suitable form to policy makers, researchers and the general public for statistical purposes. The release of such information may have the undesirable effect that information on individual entities instead of on sufficiently large groups of individuals is disclosed. The question then arises how the information available can be modified in such a way that the data released can be considered statistically useful and do not jeopardize the privacy of the entities concerned. The statistical disclosure control theory is used to solve the problem of how to publish and release as much detail in these data as possible without disclosing individual information ([6] and [7]).

This paper discusses the available tools to protect data and the option for researchers to work on-site at the Statistics Netherlands. The tables produced by Statistics Netherlands on the basis of the microdata of surveys have to be protected against the risk of disclosure. Therefore the software package τ-ARGUS [1] can be applied on the tables produced. More information about τ-ARGUS and how this package can be

J. Domingo-Ferrer (Ed.): Inference Control in Statistical Databases, LNCS 2316, pp. 203 - 212, 2002.

applied is given in section 2. Section 3 explains how microdata for research and public use microdata files can be produced using the software package μ-ARGUS [2]. The option for bona fide researchers to work on-site on richer microdata files is explained in section 4. Some examples of official statistics that have benefited from statistical disclosure control techniques are given in section 5. Finally, a discussion about the current state and some possible extensions for the ARGUS packages in section 6 concludes this paper.

2   The Release of Tables with τ-ARGUS

Many tables are produced on the basis of surveys. As these tables have to be protected against the risk of disclosure, the software package τ-ARGUS [1] can be applied. Two common strategies to protect against the risk of disclosure are table redesign and the suppression of individual values. It is necessary to suppress cell values in the tables because publication of (good approximations of) these values may lead to disclosure. These suppressions are called primary suppressions. A dominance rule is used to decide which cells have to be suppressed. This rule states that a cell is unsafe for publication if the n major contributors to that cell are responsible for at least p percent of the total cell value. The idea behind this rule is that in unsafe cells the major contributors can determine with great precision the contribution of their competitors. In τ-ARGUS the default value for n is 3 and the default value for p is 70 %, but these values can be changed easily if the user of the package prefers other values. Using the chosen dominance rule τ-ARGUS shows the user which cells are unsafe. In publications crosses (×) normally replace unsafe cell values.

As marginal totals are given as well as cell values it is necessary to suppress further cells in order to ensure that the original suppressed cell values cannot be recalculated from the marginal totals. Even if it is not possible to recalculate the suppressed cell value exactly, it is often possible to calculate it within a sufficiently small interval. In practical situations every cell value is namely non-negative and thus cannot exceed the marginal totals in the row or column. If the size of such an interval is small, then the suppressed cell can be estimated with great precision, which is of course undesirable. Therefore, it is necessary to suppress additional cells to ensure that the intervals are sufficiently large. A user has to indicate how large a sufficiently large interval should be. This interval is called the safety range and in τ-ARGUS the default safety range has a lower bound of 70 % and an upper bound of 130 % of the cell value, however it is possible for a user to change these default values at will. A user of a table cannot see if a suppression is a primary or secondary suppression: normally all suppressed cells are indicated by crosses (×). Not revealing why a cell has been suppressed helps to prevent the disclosure of information.

Preferably the secondary suppressions are executed in an optimal way, however the definition of optimal is an interesting problem. Often, the minimisation of the number of secondary suppressions is considered to be optimal. Other possibilities are to minimise the total of the suppressed values or the total number of individual contributions to the suppressed cells. The minimisation of the total of the suppressed values is of course only relevant if all cell values are non-negative. In τ-ARGUS the option of minimising the total of the suppressed values has been implemented as the default. In τ-ARGUS version 2.0 it is also possible to minimise the total number of individual contributions to the suppressed cells. If that criterion is desired a so-called cost variable that is equal to 1 for every record has to be used to execute the secondary suppressions in τ-ARGUS version 2.0. However, the option of minimising the number of secondary suppressions itself has not yet been implemented. For future versions of τ-ARGUS, the aim is to implement more options so that the different resulting groups of secondary suppressions can be compared.

If the process of secondary suppressions is directly executed on the most detailed tables available, large numbers of local suppressions will often result. Therefore, it is better to try to combine categories of the spanning (explanatory) variables. A table redesigned by collapsing strata will have a diminished number of rows or columns. If two safe cells are combined a safe cell will result. If two cells are combined when at least one is not safe it is impossible to say beforehand if the resulting cell will be safe or unsafe, but this can easily be checked afterwards by τ-ARGUS. However, the remaining cells with larger numbers of enterprises tend to protect the individual information better, which implies that the percentage of unsafe cells tends to diminish by collapsing strata. Thus, a practical strategy for the protection of a table is to start by combining rows or columns. This can be executed easily within τ-ARGUS. Small changes in the spanning variables can most easily be executed by manual editing in the recode box of τ-ARGUS, while large changes can be handled more efficiently in an externally produced recode file which can be imported into τ-ARGUS without any problem. After the completion of this redesign process, the local suppressions can be executed with τ-ARGUS given the parameters for n, p and the lower and upper bound of the safety range.

As normally many tables are produced on the basis of a survey and the software package used for the data protection is based on individual tables, there is the risk that although each table is safe, the combination of the data in these tables will disclose individual information. This may be the case when the tables have spanning and response variables in common. The current version of τ-ARGUS does not support linked tables. Although it has an option to protect such tables, this is not warranted in the current version. However, the aim is to extend τ-ARGUS in such a way that it is able to deal with an important sub-class of linked tables, namely hierarchical tables. A hierarchical table is an ordinary table with marginals, but also with additional subtotals. Hierarchical tables imply much more complex optimisation problems to be solved than single tables. Some approximation methods exist for finding optimal solutions for these problems. The extensions of τ-ARGUS will be implemented in new versions of the package that will be released in the CASC (Computational Aspects of Statistical Confidentiality) project. The CASC project is funded under the Fifth Framework Programme for Research, Technological Development and Demonstration (RTD) of the European Union.

3   The Release of Microdata for Researchers and Public Use Microdata Files with μ-ARGUS

Many users of surveys are satisfied with the safe tables released by Statistics Netherlands. However, some users require more information. For many surveys microdata for researchers are released. The software package μ-ARGUS [2] is of help in producing these microdata for researchers. For the microdata for researchers Statistics Netherlands uses the following set of rules:

1. Direct identifiers should not be released.
2. The indirect identifiers are subdivided into extremely identifying variables, very identifying variables and identifying variables. Only direct regional variables are considered to be extremely identifying. Each combination of values of an extremely identifying variable, a very identifying variable and an identifying variable should occur at least 100 times in the population.
3. The maximum level of detail for occupation, firm and level of education is determined by the most detailed direct regional variable. This rule does not replace rule 2, but is instead an extension of that rule.
4. A region that can be distinguished in the microdata should contain at least 10 000 inhabitants.
5. If the microdata concern panel data direct regional data should not be released. This rule prevents the disclosure of individual information by using the panel character of the microdata.

In the case of most Statistics Netherlands' business statistics the responding enterprises are obliged by a law on official statistics to provide their data to Statistics Netherlands. This law dates back to 1936 and was renewed in 1996 without changing the obligation of enterprises to respond. No individual information may be disclosed when the results of these business surveys are published. The law states that no microdata for research may be released from these surveys. Statistics Netherlands can therefore provide two kinds of information from these surveys: tables and public use microdata files. Public use microdata files contain much less detailed information than microdata for research. The software package μ-ARGUS [2] is also of help in producing public use microdata files. For the public use microdata files Statistics Netherlands uses the following set of rules:

1. The microdata must be at least one year old before they may be released.
2. Direct identifiers should not be released. Also direct regional variables, nationality, country of birth and ethnicity should not be released.
3. Only one kind of indirect regional variables (e.g. the size class of the place of residence) may be released. The combinations of values of the indirect regional variables should be sufficiently scattered, i.e. each area that can be distinguished should contain at least 200 000 persons in the target population and, moreover, should consist of municipalities from at least six of the twelve provinces in the Netherlands. The number of inhabitants of a municipality in an area that can be

distinguished should be less than 50 % of the total number of inhabitants in that area.

4. The number of identifying variables in the microdata is at most 15.
5. Sensitive variables should not be released.
6. It should be impossible to derive additional identifying information from the sampling weights.
7. At least 200 000 persons in the population should score on each value of an identifying variable.
8. At least 1 000 persons in the population should score on each value of the crossing of two identifying variables.
9. For each household from which more than one person participated in the survey we demand that the total number of households that correspond to any particular combination of values of household variables is at least five in the microdata.
10. The records of the microdata should be released in random order.

According to this set of rules the public use files are protected much more severely than the microdata for research. Note that for the microdata for research it is necessary to check certain trivariate combinations of values of identifying variables and for the public use files it is sufficient to check bivariate combinations. However, for public use files it is not allowed to release direct regional variables. When no direct regional variable is released in a microdata set for research, then only some bivariate combinations of values of identifying variables should be checked according to the statistical disclosure control rules. For the corresponding public use files all the bivariate combinations of values of identifying variables should be checked.

The software package μ-ARGUS is of help to identify and protect the unsafe combinations in the desired microdata file. Thus rule 2 for the microdata for researchers and the rules 7 and 8 for the public use microdata files can be checked with μ-ARGUS. Global recoding and local suppression are two data protection techniques used to produce safe microdata files. In the case of global recoding several categories of an identifying variable are collapsed into a single one. This technique is applied to the entire data set, not only to the unsafe part of the set, so that a uniform categorisation of each identifying variable is obtained.

If a certain identifying variable is desired in many categories, it means that other identifying variables can have fewer categories. Ideally, all identifying variables would have so few categories that no more unsafe combinations in the microdata would exist and local suppressions would not be necessary. When local suppression is applied, one or more values in an unsafe combination are suppressed, i.e. replaced by a missing value. These missing values could be imputed, but this is normally not attempted as bad imputations give misleading information to users and good imputations could lead to disclosure of the individual information of respondents. Local suppressions thus limit the possibilities of analysis, as there are no longer rectangular data files to analyse. However in practice, when producing protected microdata (microdata for researchers or public use microdata files) it is hard to limit the level of detail in the identifying variables and one often needs some local suppressions to meet the data protection criteria. Therefore, after the recoding of the identifying variables interactively with μ-ARGUS the remaining unsafe combinations have to be protected by the suppression of some of the values. The software package μ-ARGUS automatically and optimally determines the necessary local suppressions, i.e. the number of values that have to be suppressed is minimised. In this way it is possible to quickly produce microdata for researchers and public use microdata files.

Small changes in the identifying variables can be executed most easily by the manual editing in the recode box of μ-ARGUS, while large changes can be handled more efficiently in an externally produced recode file which can be imported into μ-ARGUS without any problem. After this global recoding the remaining unsafe combinations will be suppressed by μ-ARGUS to obtain protected microdata. No other protected microdata may be produced from the same data set, as the data protection measures could be circumvented by combining information. Therefore, before releasing protected microdata one has to plan carefully which variables to include in these files and how to recode the identifying variables included in the file. One can produce such a file only once.

In the field of microdata several new techniques will be investigated in the CASC (Computational Aspects of Statistical Confidentiality) project. The CASC project is funded under the Fifth RTD Framework Programme of the European Union. New methodologies like post randomisation (PRAM), micro-aggregation and noise-addition will be implemented in new versions of μ-ARGUS that will be released in the near future. This will allow for experimenting with these techniques. To measure the quality of the methods applied disclosure risk and information loss models will be implemented too.

4   Working On-Site in a Secure Area within Statistics Netherlands

Some researchers need more information than is available in the released microdata for researchers or public use microdata file. As the releasing of richer data is not allowed, it is then possible for individual researchers to perform their research on richer microdata on the premises of Statistics Netherlands. Bona fide researchers have the opportunity to work on-site in a secure area within Statistics Netherlands. Researchers can choose at will between the two locations of Statistics Netherlands: Voorburg in the west of the Netherlands and Heerlen in the south of the Netherlands. The possibility to export any information is however only possible with the permission of the responsible statistical officer. They can apply standard statistical software packages and also bring their own programmes. Like all employees of Statistics Netherlands, these people who work on-site have to swear an oath to the effect that they will not disclose the individual information of respondents [3].

The researchers who work on-site on economic data have to take the rules of Statistics Netherlands' Centre for Research of Economic Microdata (CEREM) into account. The most important rules are:

- researchers must be associated with a recognised research institute (e.g. a university);
- there must be a research proposal that conforms to current scientific standards;
- the researcher and his superior have to sign a confidentiality warrant;
- the researcher obtains only access to the data needed for his project;
- the data do not contain information on names and addresses of the enterprises;
- data related to the two most recent years will not be supplied;
- it is forbidden to let data or not safeguarded intermediate results leave the premises of Statistics Netherlands;
- all prospective publications will be screened with respect to risk of disclosure;
- all publications will be in the public domain;
- a public register contains the researcher's name(s), the research project, the publication(s) and the databases provided.

The facility is not free of charge. As a rule the researcher has to pay the cost for the supply of the required data. In addition, there is a tariff for using the on-site facility.

5   Examples of Official Statistics

In September 2000 Statistics Netherlands introduced a new organisation structure. Most data are now produced in the Divisions of Business Statistics and Social and Spatial Statistics.

In the Division of Business Statistics the most important surveys are the Production Statistics. Lots of tables are produced on the basis of these surveys. It is not an easy task to develop a consistent protection strategy for these tables. Some specially developed modules are used to tackle this problem. The idea is to integrate some of these modules into τ-ARGUS so that many users can profit from them. Some bona fide researchers want to perform special research projects and work on-site in a secure area within Statistics Netherlands on the microdata of the Production Statistics. For some projects these microdata have to be matched with other surveys. In those cases Statistics Netherlands does the matching and then the resulting data set without the direct identifiers can be analysed by the bona fide researchers.

In the Division of Social and Spatial Statistics many different smaller surveys are conducted. The biggest of these surveys is the Annual Survey on Employment and Earnings (ASEE). In [5] it is described how payroll data for the ASEE are collected. The ASEE data sets contain large numbers of records and a lot of information concerning earnings.

The problem is how to handle linked ASEE tables using the current version of τ-ARGUS. As all of the tables have to be protected against the risk of disclosure, the current version of τ-ARGUS is applied to three basic tables. These are far fewer tables than are published, however, many specific tables can be constructed from the protected basic tables which will automatically also be safe. What remains is how to simultaneously protect the different basic tables. As the problem of how to solve the suppression problem in an optimal way for two or more tables simultaneously is not warranted in the current version, it was necessary to find a practical protection strategy.

In practice, two complications make our data protection process for linked ASEE tables a bit more difficult. Firstly, it is not only cell values and totals that are published, but also many subtotals. Therefore, the process must be executed at the level of the basic subtable. Secondly, if there is a choice of where to put a secondary suppression cross it is considered to be superior practice to put it in a cell that was also suppressed the previous year. Otherwise, each year a basic subtable may be safe, but the combination of such tables from consecutive years could lead to the disclosure of individual information. Many cell values do not differ substantially from year to year and often the main contributors to these cells are the same, thus good estimates can be made for suppressed cell values if the same cell is not suppressed the year before or the year after.

Currently the ASEE data set contains about 50 % of all employees in the Netherlands. The challenge is to enlarge the number of records of earnings information to all employees in the Netherlands within the next year. To reach this aim it will be of great help to use information from the Insured Persons Register, which contains a large number of records and in which the private sector is very well represented. A disadvantage of this register is that the number of variables is smaller than in the ASEE, but imputation techniques (see e.g. [4]) help to overcome this problem. Of course the enlarged survey gives new challenges in the field of statistical disclosure control. Another challenging source that has recently come available for Statistics Netherlands is the Fiscal Database. This database contains more actual information than the Insured Persons Register, but we have to be very careful about which information can be published from this source because of the strict statistical disclosure control policy with respect to this source.

Another interesting recent development is the production of matrices (aggregated microdata) that are published in Statline. Statline is a product of Statistics Netherlands to view the data on a user-friendly way and to give users the possibility to let them make their favourite tables. As users of Statline can produce any table from a matrix at will one must be careful what kind of information is included in these matrices. The number (of categories) of identifying variables per matrix can therefore only be limited and rounding is used to further protect the individual sensitive information. Such matrices are currently being produced for social security statistics, education statistics and labour statistics.

6   Discussion

The software packages τ-ARGUS and μ-ARGUS have emerged from the Statistical Disclosure Control (SDC) project that was carried out under the Fourth RTD Framework Programme of the European Union. These software packages appear to be of great help in the practice of statistical disclosure control. Many of the protection problems of statistical data can be solved using the ARGUS packages. A few of these problems were mentioned in this paper.

The manuals ([1] and [2]) are of great help for the users of the ARGUS packages. However, there are always additional things to desire. In the case of τ-ARGUS it would be of great help if linked tables, and more in particular hierarchical tables, could be dealt with in a more automated way. This need has been recognised for some time; in fact a preliminary implementation of a linked table option is already available in the current version of τ-ARGUS. A dedicated computer program (using the optimisation DLL of τ-ARGUS) has been developed by the Methods and informatics Department of Statistics Netherlands to deal with hierarchical tables. This computer program, written in C++, is called HiTaS and is a stand-alone program that will be incorporated into τ-ARGUS as part of the CASC (Computational Aspects of Statistical Confidentiality) project. The CASC project is funded under the Fifth RTD Framework Programme of the European Union.

More research is also needed into how consecutive years of the same survey can be protected from disclosure. Finally, it would be good to have more options available on how to execute the secondary suppressions. In the case of μ-ARGUS, it is important to clarify in the package the difference between protecting microdata for research and protecting public use microdata files. As μ-ARGUS can be used with lots of different protection criteria, it is important to help the users to understand how different strategies can be executed using the package. Recently, research has been directed at a perturbation method by adding stochastic noise to microdata. It would be good to have an option in μ-ARGUS to perturb data as a protection technique.

It can be concluded that there is still a lot of research to be done in the field of statistical disclosure control. Hopefully, new versions of the ARGUS packages (that include results of the on-going research) will soon be released to the user community. The production of these new versions is part of the CASC project. To promote the results of the statistical projects under the Fourth RTD Framework Programme of the European Union the AMRADS (Accompanying Measures in Research And Development in Statistics) project is funded under the Fifth RTD Framework Programme. Many courses and conferences will be organised, among other topics, about statistical disclosure control. These activities will stimulate the progress in the implementation of statistical disclosure control methods and techniques in both EU countries and transition countries.

References

1. Hundepool, A.J., Willenborg, L.C.R.J., Van Gemerden, L., Wessels, A., Fischetti, M., Salazar, J.J. and Caprara, A. (1998a), *τ-ARGUS, user's manual*, version 2.0.
2. Hundepool, A.J., Willenborg, L.C.R.J., Wessels, A., Van Gemerden, L., Tiourine, S. and Hurkens, C. (1998b), *μ-ARGUS, user's manual*, version 3.0.
3. Kooiman, P., Nobel, J.R. and Willenborg, L.C.R.J. (1999), 'Statistical data protection at Statistics Netherlands' in *Netherlands Official Statistics*, Volume 14, spring 1999, pp. 21-25.
4. Schulte Nordholt, E. (1998), 'Imputation: methods, simulation experiments and practical examples' in *International Statistical Review*, Volume 66, Nr. 2, pp. 157-180.

5. Schulte Nordholt, E. (2000), 'Statistical disclosure control of the Statistics Netherlands employment and earnings data' in *Statistical Data Confidentiality, Proceedings of the Joint Eurostat/UN-ECE Work session on Statistical Data Confidentiality held in Thessaloniki in March 1999*, European Communities, 1999, pp. 3-13.
6. Willenborg, L.C.R.J. and De Waal, A.G. (1996), 'Statistical disclosure control in practice', *Lecture Notes in Statistics 111*, Springer-Verlag, New York.
7. Willenborg, L.C.R.J. and De Waal, A.G. (2001), 'Elements of statistical disclosure control', *Lecture Notes in Statistics 155*, Springer-Verlag, New York.

Empirical Evidences on Protecting Population Uniqueness at Idescat*

Julià Urrutia and Enric Ripoll

Statistical Institute of Catalonia (Idescat)

Abstract. This paper describes the process of statistical disclosure analysis and control applied by the Statistical Institute of Catalonia (Idescat) to microdata samples from census/surveys with some population uniques. Since 1995, by means of models which allows calculation of the risk and data protection procedures, some empirical evidences have been achieved in order to check the performance of μ-ARGUS in a real situation of unique populations, with large files and re-identification keys. The analyzing way used preferably is the measuring of dimensions (to a maximum of six) and the recodification on the changes of information loss versus the disclosure risk variations in the dissemination of anonymized registers. These results should be systematically extented to both social and business (micro)data so as they could allow us to test the effectiveness of new μ-ARGUS features in the undergoing CASC project.

Keywords: Disclosure risk, information loss, unique populations, anonymized records, re-identification and subsample methods.

1 Introduction

The advances of the methodology guided to the preservation of the statistical secret, specially intense in the decade of the nineties, have not always gone accompanied by the appropriate empirical contrasts of its eventual power on their main beneficiaries side: the official statistical offices. This necessity has been accentuating as these techniques are diversified in a growing way and, mainly, in view of the absence of a conceptual framework (including the own definition of statistical disclosure, what they are sensitive data or the measure of the loss of information) and an empirical one (oneself group of data qualitative and /or quantitative) stable enough.

The difficulties are also enlarged when articulating an integral and coherent policy in this field on the statistical institutes' side. This way, the analytic power of the users of data, a software on statistical disclosure control not very capable still for agile and complete processing of big volumes of data (interrelated) and

* This work was partly subsidized by the 5th Framework program of European Commission under grant no. IST-2000-25069.

The views expressed in this paper are those of the authors and do not necessarily reflect the policies of the Statistical Institute of Catalonia (Idescat).

J. Domingo-Ferrer (Ed.): Inference Control in Statistical Databases, LNCS 2316, pp. 213–229, 2002.

the problem of the protection of the accessible databases (necessarily) from Internet represent substantial challenges to continue maintaining the simultaneous trust of offerers and users of information.

In this context, the objective of this paper is double: to point out the trajectory of the solutions rehearsed and/or adopted by the Statistical Institute of Catalonia (from now on, Idescat) in the environment of the microdata diffusion and also to present some of the results in the validation of those approaches that have been revealed more promising. For all of it, we count on the experience accumulated from 1995 around effective levels of microdata protection and the corresponding loss of information (to the statistical effects) whose position has counted on a systematic analysis of the main variables (for the most part demographic and social) from the 1991 Population Census of Catalonia.

This way, the most complete whole in the tasks on statistical disclosure control done by the Idescat, that are commented later on, corresponds to a stable but concrete scenario and, therefore, limited: the treatment of physical individuals' sociodemographic variables generally associated to an exhaustive operation of basic characteristics of the population -census- and of mainly qualitative or categorical nature.

In this sense, the document is structured in four parts. In section 1 the nature of the main investigations carried out by the Idescat are exposed, from its creation in 1990, about the microdata cession previous to the adoption of the SDC pattern using μ-ARGUS. It also includes a reference to the works on the macrodata controlled dissemination based on the random perturbations with compensation and the experimental use of the privacy homomorphisms for microdata statistical confidentiality developed outside the Idescat.

Section 2: the empirical mark of the re-identification pattern adopted by the Idescat is described to evaluate the probability of risk of statistical revelation by means of the use of the available options in μ-ARGUS (local suppressions), with which, in 1999, superior results were already obtained to the strict use of the re-identification models.

Section 3: the treatment of the loss of information associated to the procedures that offer smaller risks of statistical revelation in the dissemination of microdata files is approached, contrasting the trade-off between both concepts as the level of dimensions and thresholds of the included variables increase, and analyzing its evolution and its profile in particular.

Section 4: contains some final remarks and comments on the last two sections of this paper, which benefit of the "stability" of the conceptual and empirical framework that the Idescat has maintained in its rehearsals about microdata protection, with independence of its intrinsic value and the limitations of its eventual extrapolation and validity for other scenarios.

2   Approaches to the Macrodata and Microdata Statistical Control

Although the central purpose of this paper is limited to the controlled dissemination of microdata files through the adoption of a concrete strategy, it is worthwhile to highlight two experiences relatively innovative examined or promoted by the Idescat between 1993 and 1996: the adaptation of new masking procedures for contingency tables and the potentiality of the cryptographic techniques, fully emergent in the electronic exchanges of information.

2.1   Random Perturbations by Compensation in Macrodata on Population Census

The first of the previous experiences rehearsed by the Idescat refers to the introduction of random perturbations by compensating developed by Appel and Hoffman[1], intending an implementation of this focus that simplified the original proposal. Starting from its presentation in the first "International Seminar on Statistical Confidentiality 1992" held by ISI-Eurostat in Dublin, the work of Turmo[2] evidenced that, in the case of the variables "Place of birth" and "Academic level" for all the municipalities of Catalonia, two refinements could be incorporated: to previously treat the unique populations in the environment row/column and to adjust the marginal totals of the tables through compensating methods.

This first approach to the statistical disclosure control on tabular data was clearly placed in the segment of well-known procedures as of "modification" of confidential data and the disclosure risk of this modified method was irrelevant and it kept to high level of information (also users allowed to work with methods they are used to).

Despite the kindness and the effectiveness of the procedures applied to a set of statistical tables from 1991 Population Census of Catalonia, their applicability was committed by the concurrence of diverse elements. On one hand, the reticences on an undesirable distortion of the data original value from the users, aware of the existence of alternative and equally valid procedures guided to the "reduction" of information instead of its "modification". On the other hand, the non-existence of an appropriate computing support by the middle nineties, for the agile and powerful processing of the information that should be tabulated (including the case of hierarchical and/or linked tables) frustrate good part of the automated approaches.

Finally, the progressive orientation of the Idescat towards the articulation of the control politics of the statistical revelation in the microdata cession is justified in two elements:

- The individual records are potentially the more requested type of data by the users and/or researchers due to the growing power of their treatment teams and analysis of the primary information. This tendency is accentuated particularly in the case of data with reduced geographical levels (i.e., statistical information on European regions).

- A significant proportion of the applicable "solutions" for the preservation of the microdata statistical confidentiality is expandable to the protection of tabulated data. On the other hand, some research lines are centered in the possibility of disseminating tables of results based on a previous treatment of the individual records that tabulate in two or more dimensions (either by means of their reduction or modification).

This way, in the Idescat, procedures of statistical data protection on macrodata were being settled down based on the focus principles in favor of the reduction of the confidential information, instead of subjecting it to the opportune "modifications"[1].

2.2   Use of Privacy Homomorphisms for Macro/Microdata Statistical Confidentiality

In the field of statistical data protection for macro/microdata it is necessary to point out a second relevant experience, consisting of the development of prototypes for delegation and computing data based on cryptographic techniques. The research carried out by the team of J. Domingo-Ferrer[3] allowed to evidence the possibility that the classified level (i.e. statistical institute) can recover exact statistics from statistics obtained at an unclassified level (i.e. subcontracted) on disclosure-protected dates by means of privacy homomorphisms (PHs) for multilevel processing of classified statistical data.

Although the application field of the cryptographic techniques can extend to the multiple steps of the statistical data protection -data collection, data processing or data dissemination-, like is showed in Domingo, Mateo and Sánchez[4], the emphasis and the contrasts of its use by the Idescat were centered in their benefits in data processing. The experiences carried out in this last field starting from 1996 were of great utility for the treatment of specially sensitive data (e.g. individualized data of health).

At the same time, some practical limitations of the developed prototype were revealed and they derived from two facts:

- Strong restrictions on the analytic or computing capacity on the part of an unclassified level as far as doesn't allow to make bivariant statistical analysis (or of superior dimension) of the corresponding variables in a direct way.

---

[1] In general terms, starting from the pointed out empiric evidences, the procedures on statistical data protection applied to tabular data at the Idescat concentrate on the following rules: a) rule of the minimum frequency or minimum value of a cell (all the cells below this limit are considered confidentials). b) the dominance rule (or concentration rule) that establishes that a cell is dangerous if a minimum number of individuals contribute in more than a certain percentage of the total of the cell (it is considered as a case of predominance the one that allows to one of these individuals to deduce the value corresponding to the rest of the contributers of the cell). On the other hand, the techniques applied for the elimination of confidential cells are "primary and secondary cell suppression" and changes in the outline of the table (reduction of the table dimension or recoding of its features).

- The scarce acceptance by the potential beneficiaries due, among other reasons, to the relative technical complexity that supposes the management and maintenance of this kind of platforms.

2.3   Re-identification Model for the Release of Microdata (Sub)samples

The growing microdata demand from the users on statistics and the commitment that the offices of official statistic have of giving the available information, made that the Idescat began a cession policy of these data mainly to the research centers of the Catalan universities. Evidently the cession of data should be carried out maintaining the commitment of preservation of statistical secret and therefore, controlling the disclosure risk through some of the well-known techniques. The Idescat opted to use reduction techniques due to the previous experiences in the field of the masking techniques mentioned and to the possibility of appearance of distortions not wanted in the ceded data.

This way, in 1995 the Idescat made out a microdata file with an acceptable disclosure risk level by means of the following process:

- The file contained a sample with a sample fraction of 0.4%, corresponding to more than 245.000 records, of the 1991 Population Census of Catalonia.
- Modification of the level of aggregation of the geographical and conceptual variables as instruments of control of the disclosure risk instead of using data distortion or alteration methods.
- Calculation of the disclosure risk by means of some appropriate pattern.

Once elaborated the file and before beginning their dissemination, the Idescat should make sure that the disclosure probability was small enough as to guarantee the population's privacy.

A model based on the frequency analysis was chosen, starting from the contained data in the sample, where the selection of the useful variables for the identification would be crucial in the final analysis.

Measuring the disclosure risk required the calculation of an identification probability serial based on the frequencies of unique populations in the population and in the sample. Starting from the existent methodological literature on the topic, a method based in a subsampling technique[2] presented by Zayatz[5] was chosen.

Finally, with the encouraging results of the SDC project from the 4th Framework of European Union and, especially, the readiness of the first versions of the software packages μ-Argus and τ-Argus, the Idescat was almost forced to reconsider its methodological strategy of the statistical disclosure control for the case of micro and macrodata.

---

[2] This method is based on the use of a subsampling obtained from the sample of census individual records. The idea is to obtain an estimator of the proportion of unique cases in the population, starting from the empiric data given by the analysis of the observations behavior of the subsampling regarding the sample.

This way, in 1998, the Idescat began to work with Argus, comparing the model used by Idescat for the microdata protection process and the methodological approach built in μ-Argus in order to make it possible to evaluate the results and plan for further operations in the field of statistical disclosure control. This comparison enables us to test the performance of μ-Argus from two perspectives, to check the effectiveness of μ-Argus in a real situation of unique populations, with large files and re-identification keys, and to carry out a comparative analysis of the results obtained using μ-Argus on the Idescat sample, by considering the protection criteria applied, the level of the information lost and the resulting risk of disclosure. These results will be shown in the next section of this paper.

3   Advances and Evidences in the Sure Processing of Microdata

3.1   Empirical Framework

The aim of this part is to present the results obtained when μ-Argus has been applied to a sample of individual records from Population of Catalonia in 1991 and to extend the analysis reported by Garín and Ripoll[7] in 1999. The sample size is 245.288 records corresponding to a sample fraction of approximately 0.04, and it has been disseminating between different universities and research centers applying previously a statistical disclosure control based on re-identification techniques.

But not all the variables of the file have been introduced in the analysis, only eight of them, those that have been considered as with more identification power are part of the analyzed file. These variables are: Place of Residence, Place of Birth, Age Strata, Sex, Marital Status, Profession, Academic Level and Activity Situation.

It's necessary to consider that the file contains only qualitative data and that, therefore, the analysis level is considerably limited since any quantitative variable is not included. But to keep on working with the same microdata has been decided, on one hand to be able to carry out comparative valuations of the results with the μ-Argus application and on the other hand to be able to enlarge the results obtained previously.

The results obtained in 1999 showed that, indeed, the risk of statistical disclosure decreases with μ-Argus after the application of suppression techniques, in exchange for a loss of information. Now, it is sought to analyze, among other questions, how this loss of information is distributed.

The carried out empirical experiences[3] are guided to compare, in different ways, the changes produced in the loss of information done by μ-Argus. In the first place, the variation in the number of suppressions after applying the global recoding is analyzed, that is to say, in function of the codification used in the two main variables. On the other hand, it will be interesting to observe what

---

[3] It's been used μ-Argus version 2.5 due to problems with version 3.0 installation.

happens when the number of dimensions or of combinations of tables is increased following the information introduced in the identificative level (3, 4, 5 and 6 dimensions). On the other hand, with the increment of the threshold (threshold=1, threshold=2). Finally, the loss of information is analyzed from the analysis of the change in the distributions of the variables through a brief analysis of homogeneity for the cases of 3 and 4 dimensions.

3.2   Recoding of Variables

We would like to comment that, like in the previous works, the techniques of statistical disclosure control used have been, basically, the global recoding of variables and the local suppression. Also, two of the eight key variables contained in the file, (Place of Residence, Age Strata) are subjected to recode:

Place of Residence: Two alternative recoding sets:
1. aggregation in four categories corresponding to broad administrative divisions
2. aggregation in sixteen categories corresponding to groups of local counties with low distribution variance

Age Strata: we know that one-year stratas are dangerous, therefore we established two aggregation levels:
1. categories of 5 years
2. categories of 10 years
The combination of these variables produces the following four files:

	Age Strata	
Place of Residence	5 years strata	10 years strata
4 categories	File 4_5	File 4_10
16 categories	File 16_5	File 16_10

3.3   The Identification Level and Threshold

The most revealing variables from the data matrix have been chosen because the disclosure risk must be controlled through the analysis of these possible combinations and their identification levels. To make a decision, the following aspects have been taken into consideration:

- Experiences in previous similar operations
- The type and contents of local files that contain individual data and are accessible to the public
- The quality of the information regarding the current value of some variables, the reliability of the matching of codified data, etc.

On the other hand, it has been already mentioned that this analysis contains a 3 to 6 dimensions comparison. So, in the sixth column of the metafile, we have defined an identification level distribution for each dimension. We also should establish that, if we want to compare the results between dimensions, the distribution of the identification levels mustn't change so much. So we only change the identification level of four variables in order to obtain the fourth, fifth and sixth dimension.

Remember that, to the highest identification level corresponds the smallest value and that μ-Argus always requires that the identificative levels are correlative:

3 dimensions	IL	4 dimensions	IL
Place of Residence	1	Place of Residence	1
Place of Birth	2	Place of Birth	2
Age Strata	1	Age Strata	1
Sex	3	**Sex**	**4**
Marital Status	3	Marital Status	3
Profession	2	Profession	2
Academic Level	2	Academic Level	2
Activity Situation	3	Activity Situation	3

5 dimensions	IL	6 dimensions	IL
Place of Residence	1	Place of Residence	1
Place of Birth	2	Place of Birth	2
Age Strata	1	Age Strata	1
Sex	**5**	**Sex**	**6**
Marital Status	**4**	**Marital Status**	**5**
Profession	2	Profession	2
Academic Level	2	**Academic Level**	**3**
Activity Situation	3	**Activity Situation**	**4**

Where in bold appear the variables that have changed their identification level in relation to the previous combination of dimensions.

It has been necessary to force a change in the variable Academic Level to maintain an identification level similar to 3 since the original metafile (3 dimensions) presents only three variables with an identification level equal to 3. For this reason, the increment in the last combination of dimensions presents the change in four variables. It could be noted that the two variables subjected to recode and which produce the different four files to analyze are those that have a superior identification level (IL=1).

The threshold is the other big decision that the controller should take in order to obtain a safety file. In this paper we will compare the increment of suppressions when we increase the threshold from 1 to 2.

4 Empirical Evidences: The Distribution of the Information Loss

4.1 Number of Suppressions

The following tables show the number of suppressions that the program μ-Argus carries out to obtain a safe file and how this number varies in function of the used file (according to codification), the number of dimensions and the threshold value used.

Table 1. Number of suppressions

File 4_10	3 dimensions	4 dimensions	5 dimensions	6 dimensions
Threshold = 1	930	6016	12480	16179
Threshold = 2	1894	10230	19666	25313

File 4_5	3 dimensions	4 dimensions	5 dimensions	6 dimensions
Threshold = 1	1631	9548	18489	23010
Threshold = 2	3343	15953	28530	35324

File 16_10	3 dimensions	4 dimensions	5 dimensions	6 dimensions
Threshold = 1	2330	13563	26480	33293
Threshold = 2	4709	21963	39642	49736

File 16_5	3 dimensions	4 dimensions	5 dimensions	6 dimensions
Threshold = 1	3415	19909	37115	45157
Threshold = 2	6901	32123	55124	66893

Obviously, the number of suppressions increases with the number of dimensions as well as with the increment in the threshold value. On the other hand, the following tables are arranged of smaller to more number of suppressions according to the analyzed file.

This way, it's noted that the file previously surer is the File 4_10 and the one that needs more suppressions, that is to say, the less sure file, is the File 16_5. This aspect works together with the number of categories corresponding to the two recoded variables. The number of combined categories of the variable Place of Residence and Age Strata in the four files is:

File 4_10   15
File 4_5   25
File 16_10   27
File 16_5   37

In the analyzed cases, the number of suppressions increases when increases the level of detail of the microdata.

It could also be seen that the variable Place of Residence has bigger importance in the increment of the number of suppressions because the two files with a more added stratification regarding this variable are located in the first two places of the ordination.

But the number of suppressions doesn't agree with the number of records affected by some suppression since there are records that have suffered more than one local suppression. The fact of suppressing more than one variable for each record favors the fact that a bigger percentage of records remain intact but, on the other hand, is neither advisable to have an excess of records with lack of information in more than one variable. In this sense, Table 2 shows the percentage of records with more than one local suppression.

Table 2. Percentage of records with more than one suppression

		dim 3	dim 4	dim 5	dim 6
File 4_10	threshold 1	0.11	1.28	2.49	2.68
	threshold 2	0.37	2.07	4.02	4.66
File 4_5	threshold 1	0.18	1.51	2.87	3.17
	threshold 2	0.48	2.12	4.95	5.80
File 16_10	threshold 1	0.39	2.12	3.54	3.78
	threshold 2	0.71	3.52	5.78	6.53
File 16_5	threshold 1	0.71	2.71	4.48	4.72
	threshold 2	1.29	4.74	7.40	8.33

It could be seen how the increment in the detail level of the information in the number of dimensions and in the number of the threshold value, increases the percentage of records with more than one local suppression reaching relatively high values in some cases.

It could be also pointed out that in the analysis to 3 dimensions, the maximum number of record local suppressions is 2, and starting from the dimension 4, the maximum number of record local suppressions is 3. On the other hand, it has not been noted in any case that more than 3 variables in a record have been suppressed.

4.2   Distribution of the Suppressions by Variables

Another way to analyze the number of local suppressions is from the point of view of how this suppressions change in each one of the eight variables. In this sense, Table 3 contains the ratio of variation suppressions when increasing one dimension. The table is shown only for one of the four files analyzed since almost the same behavior pattern in all of them repeats.

It could be noted that in the step from 3 to 4 dimensions the variables Place of Birth, Sex Marital and Status possess a very superior ratio than the other ones. It is also observed how the ratios are equaled in the step from 5 to 6 dimensions. The same behavior repeats when we work with a threshold=2.

Table 3. Variation ratio in the variables by dimension changing

File 4_10	3 to 4	4 to 5	5 to 6
Place of Residence	4.27	3.78	1.26
Place of Birth	25.81	2.76	1.21
Age Strata	3.03	2.07	1.10
Sex	46.75	4.82	1.96
Marital Status	13.30	1.60	1.11
Profession	2.73	1.54	1.10
Academic Level	5.67	2.41	1.32
Activity Situation	4.21	1.27	1.06

Another question to think about is how the suppressions are distributed among the variables, that is to say, which variables suffer more suppressions. The following table shows this aspect through the percentage of suppressions regarding the total.

Table 4. Variation ratio in the variables by dimension levels

File 4_10	3 dim	4 dim	5 dim	6 dim
Place of Residence	27.6	21.1	33.2	32.3
Place of Birth	2.9	13.4	15.4	14.4
Age Strata	3.2	1.7	1.5	1.3
Sex	0.9	7.2	14.4	21.8
Marital Status	8.8	21.0	14.0	12.0
Profession	28.0	13.6	8.8	7.4
Academic Level	1.9	2.0	2.0	2.0
Activity Situation	26.7	20.1	10.6	8.7

When working with 3 dimensions, we see how the 80% of the suppressions concentrate on the variables Place of Residence, Status Profession and Activity Situation.

A singularity must be highlighted: the variables Profession and Activity Situation reduce their weight in the number of suppressions as the number of dimensions is increased, while the variable Sex suffers the opposing phenomenon. It increases the percentage of suppressions regarding the total as advances in the number of analyzed dimensions are carried out. This behavior pattern repeats in the four files and when we work with a threshold=1 or threshold=2.

Only another one more aspect to highlight. In the File 16_10 the percentage of suppressions regarding the total in the variable Place of Residence decreases until values inferior to 5%, although, intuitively, we could think that with an increment in the variable detail an increment in the number of suppressions would take place.

4.3   Increase of Suppressions in Dimension and Threshold Values Changing

It is obvious that with the increment in the number of dimension combinations an increment of the number of local suppressions takes place because the number of combinations analyzed is bigger as well as the identification risk. The same effect takes place with an increment in the threshold value because the records in danger of being not identified are not only those which are unique in the population. It's enough that only two records were equal (threshold=2) to mark this records as conflicting.

In this section it is analyzed, by means of increment ratios and for each one of the four files, how the loss of information or number of local suppressions vary *when we increase the dimensions* from 3 to 6 on one hand and from 1 to 2 in the threshold value. The ratio built is simply the quotient between two numbers of local suppressions to be able to see by which value has been multiplied the suppression number as the Table 5 shows.

Table 5. Ratio variation in dimension changing with threshold=1 and threshold=2

Threshold = 1	3 to 4	4 to 5	5 to 6
File 4_10	6.5	2.07	1.3
File 4_5	5.9	1.94	1.2
File 16_10	5.8	1.95	1.3
File 16_5	5.8	1.86	1.2

Threshold = 2	3 to 4	4 to 5	5 to 6
File 4_10	5.4	1.9	1.3
File 4_5	4.8	1.8	1.2
File 16_10	4.7	1.8	1.3
File 16_5	4.7	1.7	1.2

By observing the previous tables it seems interesting to highlight:

- A very high ratio is produced in the step from 3 to 4 dimensions, but in the following step, from 4 to 5 dimensions, this ratio falls to practically the third part and it continues falling, although in smaller measure, in the step from 5 to 6 dimensions where the ratio for each one of the four files is equal led. That is to say, as we advance in the number of analyzed combinations, the increment of suppressions in ratio terms spreads to be constant.
- When we increase the threshold value from 1 to 2, it is observed that from 3 to 4 dimensions the ratio has decreased in a constant similar to 1.1, of 4 to

5 dimensions the ratio decreases in a constant of 0.15 and in the step from 5 to 6 dimensions the ratio stays unalterable.

These results allow figuring the number of suppressions that will take place when applying μ-Argus with an increment of dimensions or of the threshold value. That is to say, if the ratios were known, the program could be applied only for a combination of dimensions equal to 3 and a threshold value equal to 1 and to estimate the changes that would take place with more dimensions or higher threshold values.

Now we will measure in Table 6 the changes in the number of suppressions in each one of the *files when passing from a threshold=1 to a threshold=2*. That is to say, for each file, it will be analyzed for each fixed level of dimensions, how the fact of increasing a unit the threshold value affects the result.

Table 6. Ratio variation in threshold changing with dimension levels

	3 dim	4 dim	5 dim	6 dim
File 4_10	2.04	1.70	1.58	1.56
File 4_5	2.05	1.67	1.54	1.54
File 16_10	2.02	1.62	1.50	1.49
File 16_5	2.02	1.61	1.49	1.48

It is quickly observed how, inside each dimension level, when passing from a threshold value 1 to 2, the ratio doesn't almost vary with the used code level. This way, when working with a dimension level 3, the number of suppressions duplicates in the four files. With a dimension level equal to 4, it multiplies by approximately 1.65 and in higher levels also takes place a new interesting fact: the ratio stays with what seems to be a limit located in approximately 1.5.

Summing up, these results seem to indicate that when the threshold value is increased, the quotient among the number of suppressions tends to not depending on the number of increasing changing dimensions.

4.4 Disclosure Risk Decreasing

Two kinds of analysis have been made to measure how the disclosure risk decreases when we obtain a safe file by means of μ-Argus. In fact, it is an approaching attempt to the protective effectiveness of μ-Argus software by calculating the decrease in the disclosure risk. For this reason, firstly, the results have been compared on the four files applying a local suppression in 3 and 4 dimensions and applying the calculation of the disclosure risk through re-identification techniques.

Evidently the identification risk decreases as more combinations of variables are analyzed but in exchange for increasing the level of lost information. A considerable decrease on disclosure risk takes place when applying a dimension increment (see Table 7).

Table 7. Disclosure risk decreasing in dimension changing from 3 to 4

	%
File 4_10	32.09
File 4_5	31.10
File 16_10	29.45
File 16_5	39.46

On the other hand, an experiment has been carried out consisting on obtaining two independent samples (10,000 records) from our original file with the same sampling fraction (0.404) that the one that represents this file regarding the total population, as we have seen in section 2. Later on, to one of the samples, the program μ-Argus has been applied with a threshold equal to 1 and with combinations of 4 dimensions. After calculating the disclosure risk in the two samples, the decrease of this risk has been measured, as it could be seen in Table 8.

Table 8. Disclosure risk decreasing with vs without μ-Argus performance (threshold=1, dim=4)

	%
File 4_10	66.98
File 4_5	72.60
File 16_10	59.71
File 16_5	69.56

4.5   Homogeneity of the Variables Distribution

One must keep in mind that when a file with microdata is spread, what is sought is that the analysts obtain some results the most similar possible to the original information, that is to say, the variables should maintain the distribution function.

To see how the option of local suppression acts in μ-Argus a test of homogeneity has been carried out comparing the variables distributions in the original file with the distributions of the safe file produced by μ-Argus. This test has been carried out for each one of the four analysis files, with combinations of 3 and 4 dimensions, with a threshold value equal to 1 and at a level of analysis univariant.

Subsequently, two tables are presented (Tables 9 and 10) with the result of the test in terms of to reject or not to reject the variable, that is to say, if the function of variable distribution has suffered important changes or stays between some acceptable margins.

Table 9. Homogeneity test in 3 dimensions

	Homogeneity?			
	File 4_10	File 4_5	File 16_10	File 16_5
Place of Residence	Yes	Yes	Yes	Yes
Place of Birth	Yes	Yes	Yes	Yes
Age Strata	Yes	Yes	Yes	Yes
Sex	Yes	Yes	Yes	Yes
Marital Status	Yes	Yes	No	No
Profession	Yes	Yes	No	No
Academic Level	Yes	Yes	Yes	Yes
Activity Situation	Yes	Yes	Yes	No

It is observed that most of the changes are not significative, which means that the null hypothesis that the distribution function has not varied is accepted. The files with an Age Strata with 10 year-old strata don't reject any variable while the files with strata of 5 year-old age present problems in the variables Marital Status, Profession and Activity situation, this last one only in the File 16_5.

Table 10. Homogeneity test in 4 dimensions

	Homogeneity?			
	File 4_10	File 4_5	File 16_10	File 16_5
Place of Residence	No	No	Yes	No
Place of Birth	No	No	No	No
Age Strata	Yes	Yes	No	Yes
Sex	No	Yes	Yes	Yes
Marital Status	Yes	No	No	No
Profession	Yes	No	No	No
Academic Level	Yes	Yes	Yes	Yes
Activity Situation	Yes	No	No	No

In view of the previous results, we can highlight the following aspects:

1. When passing from 3 to 4 dimensions, the increment of rejected variables is considerable, one of them is rejected systematically (Place of Birth). On the other hand, the Education Level stays unalterable and the other ones depend on the file we are working with.
2. If the distribution of the variables are analyzed one to one, we can see that the variables more easily rejected are those distributed internally in a heterogeneous way and, however, the suppressions are not carried out in a similar proportion but are distributed homogeneously among the different

categories. This causes that some category maintains its weight practically intact while, in another one, it could have varied considerably.

3. It has also been observed that it is more frequent that a variable with many categories would be rejected easier than a variable with few categories, although this condition is not always completed like could be observed with the variable Education Level (21 categories) or with the variable Place of Birth (3 categories).

4. Bivariate analyzes have been carried out but in most of the cases the homogeneity test doesn't support the null hypothesis.

5. Although it has not still been tested empirically, we intuitively suppose that if the number of dimensions or the threshold value increases more, the number of rejections will continue increasing.

6. We also point out the possibility to carry out other kind of homogeneity tests for qualitative data where the variation of marginal totals would be taken into consideration, as McNemar test, for instance.

5   Global Assessments

In this paper some empirical results in the application of the software package μ-Argus have been showed, and supplement other works carried out in the Idescat. Of those results the following valuations can be extracted as the most important:

1. For the construction of the safe files, and the consequent rising decrease of the disclosure risk, the options of global recoding and local suppression have been used. A decrease of the disclosure risk has been achieved indeed and the obtained files can be considered sure enough.

2. On the other hand, with the increment in the number of dimensions and the threshold value, the quantity of lost information tends to be too high as for not keeping in mind that the safe file cannot be satisfactory. In fact, we have seen how the probability distributions of certain variables could be modified, especially the more homogeneous the suppressions would be among the categories in a heterogeneous variable.

3. Argus works under the hypothesis that the processed file refers to the global population and not to a sample of the same one. When working with samples the disclosure risk decreases in great measure and, therefore, it would maybe not be necessary to carry out a number so high of suppressions since the unique populations in the sample shouldn't be unique in the population. It would be interesting to be able to introduce a parameter that will take into account this fact.

4. It would be convenient, on one hand, more precise and extensive documentation on different options, algorithms and proceedings used by Argus and, on the other hand, some analysis procedures in the same software package to evaluate its effectiveness, like basic descriptive analysis, analysis of the suppressed data distribution, and so on.

As partner of the CASC project in the 5th Framework program of European Commission, the Idescat hope to continue testing the software package Argus in order to extend the results to both social and business data and to improve our policy and guidelines about the statistical disclosure control.

References

1. Appel, G. and Hoffman, D.J.: "Perturbation by compensation". Preproceedings of the International Seminar on Statistical Confidentiality, ISI-Eurostat. Dublin, September 1992. Final version published by Eurostat in 1993.
2. Turmo, J.: "Random perturbations by compensation: a method to protect confidential statistical information" [in Catalan]. Quaderns d'Estadística i Investigació Operativa (Qüestiió), **17**, 3, 1993, 413–435.
3. Domingo-Ferrer, J.: "Privacy homomorphisms for statistical confidentiality". Quaderns d'Estadística i Investigació Operativa (Qüestiió), **20**, 3, 1996, 505–521.
4. Domingo-Ferrer, J., Mateo, J.M., and Sánchez, R.X.: "Cryptographic techniques in statistical data protection". Joint ECE/Eurostat Work Session on Statistical Confidentiality. Thessaloniki, Greece, March 1999. Final version published by Eurostat in 1999.
5. Zayatz, L.V.: Estimation of the percent of unique population elements on a microdata file using the sample. Bureau of Census. Statistical Research Division Report Series (1991), Census/SRD/RR-91/08.
6. Garín, A.: "Control of statistical disclosure risk in dissemination of demographical microdata, applied on a sample of individual records from Catalonia's Population Census 1991" [in Catalan]. Quaderns d'Estadística i Investigació Operativa (Qüestiió), **20**, 3, 1996, 523–546.
7. Garín, A. and Ripoll, E.: "Performance of μ-Argus in disclosure control of uniqueness in populations". Joint ECE/Eurostat Work Session on Statistical Confidentiality. Thessaloniki, Greece, March 1999. Final version published by Eurostat in 1999.

Author Index

Lecture Notes in Computer Science

For information about Vols. 1–2248
please contact your bookseller or Springer-Verlag

Vol. 2284: T. Eiter, K.-D. Schewe (Eds.), Foundations of Information and Knowledge Systems. Proceedings, 2002. X, 289 pages. 2002.

Vol. 2285: H. Alt, A. Ferreira (Eds.), STACS 2002. Proceedings, 2002. XIV, 660 pages. 2002.

Vol. 2286: S. Rajsbaum (Ed.), LATIN 2002: Theoretical Informatics. Proceedings, 2002. XIII, 630 pages. 2002.

Vol. 2287: C.S. Jensen, K.G. Jeffery, J. Pokorny, Saltenis, E. Bertino, K. Böhm, M. Jarke (Eds.), Advances in Database Technology – EDBT 2002. Proceedings, 2002. XVI, 776 pages. 2002.

Vol. 2288: K. Kim (Ed.), Information Security and Cryptology – ICISC 2001. Proceedings, 2001. XIII, 457 pages. 2002.

Vol. 2289: C.J. Tomlin, M.R. Greenstreet (Eds.), Hybrid Systems: Computation and Control. Proceedings, 2002. XIII, 480 pages. 2002.

Vol. 2291: F. Crestani, M. Girolami, C.J. van Rijsbergen (Eds.), Advances in Information Retrieval. Proceedings, 2002. XIII, 363 pages. 2002.

Vol. 2292: G.B. Khosrovshahi, A. Shokoufandeh, A. Shokrollahi (Eds.), Theoretical Aspects of Computer Science. IX, 221 pages. 2002.

Vol. 2293: J. Renz, Qualitative Spatial Reasoning with Topological Information. XVI, 207 pages. 2002. (Subseries LNAI).

Vol. 2295: W. Kuich, G. Rozenberg, A. Salomaa (Eds.), Developments in Language Theory. Proceedings, 2001. IX, 389 pages. 2002.

Vol. 2296: B. Dunin-Kęplicz, E. Nawarecki (Eds.), From Theory to Practice in Multi-Agent Systems. Proceedings, 2001. IX, 341 pages. 2002. (Subseries LNAI).

Vol. 2297: R. Backhouse, R. Crole, J. Gibbons (Eds.), Algebraic and Coalgebraic Methods in the Mathematics of Program Construction. Proceedings, 2000. XIV, 387 pages. 2002.

Vol. 2299: H. Schmeck, T. Ungerer, L. Wolf (Eds.), Trends in Network and Pervasive Computing – ARCS 2002. Proceedings, 2002. XIV, 287 pages. 2002.

Vol. 2300: W. Brauer, H. Ehrig, J. Karhumäki, A. Salomaa (Eds.), Formal and Natural Computing. XXXVI, 431 pages. 2002.

Vol. 2301: A. Braquelaire, J.-O. Lachaud, A. Vialard (Eds.), Discrete Geometry for Computer Imagery. Proceedings, 2002. XI, 439 pages. 2002.

Vol. 2302: C. Schulte, Programming Constraint Services. XII, 176 pages. 2002. (Subseries LNAI).

Vol. 2303: M. Nielsen, U. Engberg (Eds.), Foundations of Software Science and Computation Structures. Proceedings, 2002. XIII, 435 pages. 2002.

Vol. 2304: R.N. Horspool (Ed.), Compiler Construction. Proceedings, 2002. XI, 343 pages. 2002.

Vol. 2305: D. Le Métayer (Ed.), Programming Languages and Systems. Proceedings, 2002. XII, 331 pages. 2002.

Vol. 2306: R.-D. Kutsche, H. Weber (Eds.), Fundamental Approaches to Software Engineering. Proceedings, 2002. XIII, 341 pages. 2002.

Vol. 2307: C. Zhang, S. Zhang, Association Rule Mining. XII, 238 pages. 2002. (Subseries LNAI).

Vol. 2308: I.P. Vlahavas, C.D. Spyropoulos (Eds.), Methods and Applications of Artificial Intelligence. Proceedings, 2002. XIV, 514 pages. 2002. (Subseries LNAI).

Vol. 2309: A. Armando (Ed.), Frontiers of Combining Systems. Proceedings, 2002. VIII, 255 pages. 2002. (Subseries LNAI).

Vol. 2310: P. Collet, C. Fonlupt, J.-K. Hao, E. Lutton, M. Schoenauer (Eds.), Artificial Evolution. Proceedings, 2001. XI, 375 pages. 2002.

Vol. 2311: D. Bustard, W. Liu, R. Sterritt (Eds.), Soft-Ware 2002: Computing in an Imperfect World. Proceedings, 2002. XI, 359 pages. 2002.

Vol. 2312: T. Arts, M. Mohnen (Eds.), Implementation of Functional Languages. Proceedings, 2001. VII, 187 pages. 2002.

Vol. 2313: C.A. Coello Coello, A. de Albornoz, L.E. Sucar, O.Cairó Battistutti (Eds.), MICAI 2002: Advances in Artificial Intelligence. Proceedings, 2002. XIII, 548 pages. 2002. (Subseries LNAI).

Vol. 2314: S.-K. Chang, Z. Chen, S.-Y. Lee (Eds.), Recent Advances in Visual Information Systems. Proceedings, 2002. XI, 323 pages. 2002.

Vol. 2315: F. Arbab, C. Talcott (Eds.), Coordination Models and Languages. Proceedings, 2002. XI, 406 pages. 2002.

Vol. 2316: J. Domingo-Ferrer (Ed.), Inference Control in Statistical Databases. VIII, 231 pages. 2002.

Vol. 2317: M. Hegarty, B. Meyer, N. Hari Narayanan (Eds.), Diagrammatic Representation and Inference. Proceedings, 2002. XIV, 362 pages. 2002. (Subseries LNAI).

Vol. 2318: D. Bošnački, S. Leue (Eds.), Model Checking Software. Proceedings, 2002. X, 259 pages. 2002.

Vol. 2319: C. Gacek (Ed.), Software Reuse: Methods, Techniques, and Tools. Proceedings, 2002. XI, 353 pages. 2002.

Vol. 2322: V. Mařík, O. Štěpánková, H. Krautwurmová, M. Luck (Eds.), Multi-Agent Systems and Applications II. Proceedings, 2001. XII, 377 pages. 2002. (Subseries LNAI).

Vol. 2324: T. Field, P.G. Harrison, J. Bradley, U. Harder (Eds.), Computer Performance Evaluation. Proceedings, 2002. XI, 349 pages. 2002.

Vol. 2329: P.M.A. Sloot, C.J.K. Tan, J.J. Dongarra, A.G. Hoekstra (Eds.), Computational Science – ICCS 2002. Proceedings, Part I. XLI, 1095 pages. 2002.

Vol. 2330: P.M.A. Sloot, C.J.K. Tan, J.J. Dongarra, A.G. Hoekstra (Eds.), Computational Science – ICCS 2002. Proceedings, Part II. XLI, 1115 pages. 2002.

Vol. 2331: P.M.A. Sloot, C.J.K. Tan, J.J. Dongarra, A.G. Hoekstra (Eds.), Computational Science – ICCS 2002. Proceedings, Part III. XLI, 1227 pages. 2002.

Vol. 2332: L. Knudsen (Ed.), Advances in Cryptology – EUROCRYPT 2002. Proceedings, 2002. XII, 547 pages. 2002.